AF555614

ॐ

IMPROVEMENTS IN EFFICIENCY OF YEASTS FOR ETHANOL PRODUCTION

Nandita Gupta

जय गुरुदेव

ISBN13: 978-93-95766-48-7 Paperback Edition
ISBN13: 978-93-95766-46-3 Hardbound Edition
ISBN13: 978-93-95766-47-0 Digital Edition

Title: **Improvements in Efficiency of Yeasts for Ethanol Production**
Author: **Nandita Gupta**

Printed and Published by
Devotees of Sri Sri Ravi Shankar Ashram
34 Sunny Enclave, Devigarh Road,
Patiala 147001, Punjab, India

https://advaita56.in/
The Art of Living Centre

24th March 2023 Matsya Avatar Jayanti, Gauri Puja, Shukla Paksha, Chaitra Navratri Tritiya tithi, Svayambhu Manu Adi tithi, Ashwini Nakshatra, Sarvartha Siddhi Yoga, Vasant Ritu, Uttarayana.

On this day in 1603 Scottish King James VI becomes King James I of England, thus Scotland and England become one nation, 1603 Tokugawa Shogunate kingship established in Japan, 1801 Tsar Alexander I takes reign of Russia, 1882 Rober Koch discovers tubercle bacillus and establishes germ theory, 1921 Women's Olympiad begins in Monte Carlo - Monaco the first international women's sports event, 1924 Greece becomes sovereign republic, 1999 Film premiere of Matrix – based on Yoga Vasistha, 2008 Bhutan's first election moving from monarchy to democracy.

Vikram Samvat 2080 Nala, Saka Era 1945 Shobhakrit

1st Edition March 2023

जय गुरुदेव

IMPROVEMENTS IN EFFICIENCY OF YEASTS FOR ETHANOL PRODUCTION

by

NANDITA PAUL

(L-91-BS-75-D)

DISSERTATION

Submitted to the Punjab Agricultural University
in partial fulfillment of the requirements
for the degree of

DOCTOR OF PHILOSOPHY

IN

MICROBIOLOGY

[Minor Field: Biochemistry]

Department of Microbiology
College of Basic Science & Humanities
PUNJAB AGRICULTURAL UNIVERSITY
LUDHIANA - 141004

1995

DEDICATED TO

The Almighty God

who has given me such loving
Parents, Brother and Sister

Specially Dedicated to

My beloved Mother KAVITA and respected Father SURINDAR PAUL AGGARWAL who are a bundle of energy and hard work, inspiring us all in all facets of life.

Photo dated 5 July 2015 12:19pm at Punjabi Bagh home Satsang.

ACKNOWLEDGEMENTS

I feel lacking in words to express my sense of profound veneration and sincere thanks that I owe my major advisor, Dr. H. K. Tewari, Mycologist, Department of Microbiology, for his indispensable suggestions, constant supervision, constructive criticism and taking keen interest for inculcating in me the spirit of self-reliance so that I could work independently. His unfeigned guidance, radiating brilliance and pose will inspire me at every step in life ahead.

My heartfelt thanks go to Dr. P. K. Khanna, Mycologist, (Mushrooms) Department of Microbiology, for his keen interest and judicious guidance during the pursuance of this study. I am also highly obliged to the other members of my advisory committee, Dr. H. S. Garcha, Senior Mycologist, Department of Microbiology, Dr. G. L. Soni, Senior Biochemist, Department of Biochemistry and Dr. J. S. Kanwar, Professor, Department of Horticulture, for evaluating my research work at every step and for providing valuable suggestions and encouragement.

My special thanks are due to Dr. S. S. Marwaha, Professor-cum-Head, Department of Biotechnology, Punjabi University, Patiala, for generously providing cultures used in present study.

I am grateful to Dr. R. K. Sedha, Professor, and Dr. (Mrs.) Neelam, Assistant Microbiologist, Department of Microbiology, for extending cooperation and necessary facilities whenever needed.

I am profoundly indebted to Dr Shammi Kapoor without whose untiring help and prompt advices I could not have completed my experimental studies in time.

I owe immensely to my parents, my ever-replenishing sources of strength and encouragement, for their endless love and sacrifices during my entire academic life. This dissertation is a meagre token of love towards them for whatever they have done for me. I have a great privilege to make a special mention of my sister Sangeeta, and brothers, Ashwini and Rochak, for their moral encouragement, support and affection which helped me all along to forge my way ahead.

Words cannot convey my deep sense of indebtedness towards Saru, Anita, Alka, Gurpreet, Neerja Di and Devinder. It was their absolute involvement in my day to day life, academic as well as personal, that made this work possible. My heartfelt appreciation goes to Rupinder, Maninder and Nikunj for boosting me and for being there when needed. I shall always miss the cheerful company and help of Jagminder, Harvir, Neelam, Neerja, Harkirat, Parveen Di, Sandhu, Seema, Ratna & Aman.

My sincere gratitude is due to Mrs. Balwinder Kaur, Lab Assistant, for her cooperation and sense of belonging. Special recognition goes to Miss. Savita Sharma, for laser typesetting the dissertation with utmost care.

Financial assistance in the form of merit fellowship provided by Punjab Agricultural University, during the Ph.D. tenure is gratefully acknowledged.

Dated: 9th Aug '95 [Nandita Paul]

CERTIFICATE I

This is to certify that the dissertation entitled, "Improvements in Efficiency of Yeasts for Ethanol Production", submitted for the degree of Doctor of Philosophy in the subject of Microbiology [Minor: Biochemistry] of the Punjab Agricultural University, Ludhiana, is a bonafide research work carried out by Ms. Nandita Paul [L-91-BS-75-D] under my supervision and that no part of this dissertation has been submitted for any other degree.

The assistance and help received during the course of investigation have been fully acknowledged.

[Dr. H. K. Tewari]
Major Advisor, Mycologist,
Deptt. of Microbiology,
Punjab Agricultural University, Ludhiana

CERTIFICATE II

This is to certify that the dissertation entitled " Improvements in Efficiency of Yeasts for Ethanol Production", submitted by Ms. Nandita Paul (L-91-BS-75-D) to the Punjab Agricultural University, in partial fulfilment of the requirements for the degree of Doctor of Philosophy in the subject of Microbiology [Minor: Biochemistry] has been approved by the Student's Advisory Committee after an oral examination on the same, in collaboration with an external examiner.

3.11.1995

Major Advisor
[Dr. H. K. Tewari]

External Examiner
[Dr. J. K. Gupta], Professor
Deptt. of Microbiology
Punjab University, CHANDIGARH

Head of the Department
[Dr. R. S. Kahlon]

4.1.1996

Dean, Post-graduate Studies
[Dr. K. D. Menon]

Title of Dissertation:	**Improvements in Efficiency of Yeasts for Ethanol Production**
Name of Student:	**Nandita Paul**
Admission No.:	L-91-BS-75-D
Major Subject:	**Microbiology**
Minor Subject:	Biochemistry
Name and Designation of Major Advisor:	Dr. H. K. Tewari, Mycologist
Degree to be awarded:	**Ph.D.**
Year of award of degree:	**1995**
Total pages in Dissertation:	73 + xix
Name of the University:	**Punjab Agricultural University**, Ludhiana - 141004, Punjab, India

ABSTRACT

Eight strains of yeasts, namely *S. cerevisiae-220*, *S. cerevisiae-360*, *S. cerevisiae-181/3c*, *Trichosporon beigelli*, *Candida haemulonii*, *C. parapsilosis*, *Candida sp.-3(1)* and *Candida sp.-5(2)* have been evaluated for their potential to ferment cane sugars to ethanol at three temperatures. *S. cerevisiae-220* utilized 84.0% substrate in 60 h at 30 ± 1 °C. *S. cerevisiae-360* fermented substrate more efficiently (69.3%) at 38 ± 1 °C. *S. cerevisiae-220* multiply faster at 30 ± 1 as well as at 38 ± 1 °C.

S. cerevisiae-220 could withstand 13.0% ethanol whereas *S. cerevisiae-360* tolerated only 8.0% of ethanol supplemented in the growth medium. The protoplasts of *S. cerevisiae-220* and *S. cerevisiae-360* fused together could not generate ethanol more efficiently than their parents at 38 ± 1 °C. *S. cerevisiae-220* adapted at 38 ± 1 °C utilized 64.4% of substrate during 60 h.

Wine with commercially outstanding quality has been prepared from *Zingiber officianale* with *S. cerevisiae-220-A* at 38 ± 1 °C.

Dr. H. K. Tewari Nandita Paul

TABLE OF CONTENTS

List of Figures

List of Tables

List of Equations

Abbreviations

v/v = volume/volume, w/v = weight in volume, h = hour, ml = milliliter
°C = degree Celsius, °B = degree Brix, nm = nanometer
KMS = Potassium metabisulphite,
M = mole = molar mass = moles/liter, mM = millimoles. N = normality of solution.

Equipment

Erlenmeyer Flask, Bausch and Lomb Spectronic-20 spectrophotometer, Erma hand refractometer, Elico pH meter, Haemocytometer.

Chapter

1. INTRODUCTION

The fruit and vegetable production has been steadily increasing in India and in Punjab state in particular. Improved crop production technologies, superior vegetatively propagated planting material, pest and disease control measures, intensive extension activities and incentives from the State Government have contributed a lot in fruit and vegetable production. This increased production has created serious post-harvest problems for the farmers because these commodities being perishable, half short shelf life and thus cannot be stored at room temperature like grain, cotton or other farm produce. In Punjab about 25-30 percent of fruits and 15-40 percent of vegetables are wasted annually. During the ripening period in summers, it is very difficult to store the produce even for a day at atmospheric temperature. Cold storages, no doubt, play a valuable role in preserving and regulating the supply of fruits and vegetables in the market, but the present storage capacity is too low to meet the present requirements. Surplus produce can be processed into commercially acceptable products like jams and jellies. The utilization of surplus fruits and vegetables for the production of alcohol will serve as a boon for the progressive farmers.

The production of ethanol dates back with some of man's first discoveries. Originally, it was produced by the fermentation of sugary materials and was consumed by ancient races for its intoxicating effects. Archaeological evidences indicated that alcoholic beverage formation is 1000 years old. The decline in the non-renewable fossil fuel reserves has brought about the realization to develop bioprocesses for producing ethanol from renewable resources. Ethanol is among the non-petroleum fuel options which can be produced from resources available in abundance to substitute for the use of coal and petroleum.

Saccharomyces cerevisiae is the only yeast currently being used for alcohol production on the commercial scale. Ethanol can be produced from horticultural raw and waste materials containing starch or fermentable sugars. Industrial alcohol using *S. cerevisiae,* has been reported to be produced from three types of raw materials, saccharine (sugarcane molasses etc.); starchy materials (potato, cassava whey) and cellulosics (rice straw, etc.). India, which is the largest producer of sugarcane, annually produces 2.5×10^6 tons of molasses as a by-product of sugar industry. Molasses serves as raw material for about 194 distilleries in the country with an installed capacity of over 1.13 billion liters of ethanol, but with an annual production of 900-1000 million liters per annum (Sharma and Tauro, 1986). Ethanol is used as a solvent, germicide, antifreeze, liquid fuel and above all for beverage production.

The optimum temperature range for distiller's yeast is 25 to 30°C. But in tropical and subtropical regions like India, the temperature during summer ranges between 40 – 45°C, resulting in a decrease in its fermentation efficiency. The temperature of fermenters is normally 3 – 5°C higher than the environmental temperature because 11.7 Kcal of heat is produced for every kilogram of substrate consumed. Moreover, ethanol producing efficiency is also affected by the coupled effect of high temperature and ethanol (Isabel and van Uden, 1983). To avoid premature termination of fermentation, process the fermentors are usually cooled by spraying water, which is one of the major factors adding to the cost of the product. However, fermentation using thermotolerant yeasts is associated with potential benefits like improved efficiency of cooling, ability to withstand heat-shock, and more favorable economics of high production of ethanol during fermentation. High ethanol fermentation efficiency is essential to achieve a high ethanol concentration in distillation stream to reduce steam cost and a high ethanol production to keep the fermentation capital cost low.

Experts in ethanol fermentation are interested to improve ethanol productivity by developing various techniques such as continuous fermentation (Hospodke, 1968), rapid fermentation (Nagodawithana et al., 1974), fermentation of concentrated wort (Rose, 1976), vacuum fermentation with recycling (Verma et al., 1983). However, the adoption of the techniques by existing distilleries requires a capital investment, as a result these distilleries are likely to continue the present setup for a considerable time in future.

In view of the above discussion, the present study has been undertaken to develop suitable yeast strains which can produce ethanol efficiently at elevated temperatures. The proposed research work entitled "Improvements in efficiency of yeasts for ethanol production" is aimed to fulfil the following objectives:

Objectives

1. Collection of yeast cultures.
2. Screening of yeast for ethanol fermentation efficiency (Amerine et al., 1967).
3. Temperature and ethanol tolerance profile studies of yeast.
4. Production of wine at high temperature.

Chapter

2. REVIEW OF LITERATURE

Ethanol is one of the most versatile organic compounds produced by *Saccharomyces cerevisiae* from molasses for industrial use (Patureau, 1969). Besides its potable use and as a chemical feed stock in the industry, ethanol has of late started attracting much attention as a gasoline supplement/alternate fuel. Various raw materials containing reducing sugars such as, glucose, sucrose, fructose and other invert sugars are the substrates to be metabolized by yeast for ethanol production. In developing countries, significant improvement in ethanol production technology is being explored, using non-traditional raw materials like sugarcane bagasse, cassava, cellulose etc. (Lindeman and Rocchiccioli, 1979; Carroll, 1983).

The exhaustive reserves of fossil fuels have led to a steep rise in petroleum prices all over the world especially in developing countries, like India. The supplementation of petroleum with ethanol can, therefore, help the country to reduce the oil imports, thus alleviating the stress on the country's economy. Besides the availability of basic raw materials in abundance, the requirement for efficient yeasts for ethanol generation is of paramount importance. The available literature on the improvement of yeast strain for ethanol production has been reviewed under the following headings:

- Temperature tolerance
- Ethanol tolerance
- Genetic improvement
- Protoplast studies

- Production of ethanol
- Production of wine

2.1 Temperature Tolerance

Temperature is one of the most important factors in ethanol formation as it exerts a profound effect on all aspects of the growth metabolism and survival. Higher temperature profile leads to decreased cell viability and simultaneous thermal death (Stokes, 1970).

Savackuk and Tsigankov (1971) reported that a temperature of 30-32 °C was the most favorable for ensuring good quality of rectified alcohol and higher yield. Walsh and Martin (1977) observed that optimum and maximum tolerable temperatures for growth and fermentation are strongly strain dependent. Navarro and Durand (1978) and Krouwel and Braber (1979) observed decreased growth and higher ethanol productivity at temperatures ranging between 30-39 °C. At higher temperature, yeast plasma membrane permeability increases, leading to a higher ethanol production (Kleinans *et al.*, 1979).

Thermotolerant variants with respiratory deficiency appeared in wine yeasts after exposure to higher temperatures (37 °C). In addition to reduction in respiration, they exhibited intensification in fermentative functions (Kishkovskaya and Bury'an, 1980). Lee *et al.*, (1980) reported a temperature of 34 °C optimum for growth while 37-43 °C for maximum specific ethanol production rate. At 37 °C, ethanol inhibition increased significantly while inhibitory effects of ethanol on growth and specific ethanol production rate was unaffected. Leao and van Uden (1982) reported that ethanol and other alkanols enhance thermal death in yeast. Primary and secondary heat shocks increased the ethanol and thermotolerance transiently (Watson et al., 1984). Seiko *et al.,* (1985) investigated the fermentation ability of several collected and isolated strains of yeast under conditions of low pH, high temperature, high ethanol and high sugar concentration and compared to that of *S. cerevisiae* (IFO- 0224). Barillere *et al.,* (1985) studied heat resistance of different species of *Saccharomyces* and the effect of experimental conditions on the shape of survival curves. Isabel and van Uden (1986) reported that ethanol causes ethanol induced death, rather than ethanol induced thermal death. Combination of temperature and ethanol concentration outside the temperature profile of maximum ethanol tolerance resulted in exponential death, specific rate of which increased with increasing ethanol concentration. Richter and Becker (1987) and Richter (1987) studied the effect of temperature on microbial ethanol production and gave a thermodynamic interpretation of complete temperature profile curve of ethanol formation.

Xia Gao and Fleet (1988) studied the influence of temperature and pH on the survival and growth of three wine yeasts in presence of varying ethanol concentrations. Spindler *et al.,* (1988) selected five temperature tolerant yeasts for simultaneous saccharification and fermentation on cellulose. Yamamura *et al.,* (1988) studied the effects of elevated temperature on growth, respiratory deficient mutation, respiratory activity and ethanol production in yeast. The relationship between ethanol production and flocculation at high temperature was studied by Moriya *et al.,* (1988) using *Saccharomyces cerevisiae* bred for ethanol production from beet molasses. Oderinde and Esuosa (1988) observed the effects of pH, sugar concentration and temperature on the alcoholic fermentation of Nigerian cane molasses using *S. cerevisiae* and concluded that these molasses are suitable for industrial ethanol production. Laluce *et al.,* (1989) have attempted to improve the thermotolerance of amylolytic yeasts by protoplast fusion. Petite mutants of *Saccharomyces diastaticus* and respiratory competent strain of *S. cerevisiae* which grow and ferment at 40 °C were used in the protoplast fusion procedure. D'Amore *et al.,* (1989) screened various yeasts strains for their ability to grow on glucose at 40 °C . Fleet *et al.,* (1989) showed the effect of temperature on growth and ethanol tolerance of yeasts during wine fermentation.

Calazans *et al.,* (1990), while studying the effect of glucose concentration, initial pH and temperature on the alcoholic fermentation by *Zymomonas mobilis* ZAP, observed that ethanol yield was little affected by temperature in the range 30- 37.5 °C, though the optimum was 35 °C. Saito *et al.,* (1990) recloned a cellulase gene from a thermophilic anaerobe in *Saccharomyces cerevisiae*. The maximum level of gene expression in the recombinant yeast was 4.4 times higher than that in *Escherichia coli* transformant harbouring the same plasmid. Cellulase activity was observed only within the yeast cells. Neelam and Amarjit (1991) used thermotolerant yeast and its UV resistant mutants for ethanol production at 37 °C from cane molasses. Ballesteros *et al.,* (1991) reported simultaneous saccharification and fermentation of cellulosic substrates to ethanol using *Kluyveromyces marxians* and *K. fragilis* at temperatures in the range of 32-45 °C, Growth and fermentation characteristics of new selected thermotolerant strains of *Saccharomyces* at high cell densities were studied by Laluce *et al.,* (1991).

Groot *et al.,* (1992) studied the effect of repeated temperature shock on baker's yeast and observed that a temperature increase from 30 °C in the fermenter to 35 °C or more in the recycle loop led to a significantly lower ethanol concentration in the broth. Kida *et al.,* (1992) constructed a thermotolerant flocculating yeast strain *S. cerevisiae* KF-7 by protoplast fusion of the flocculating strain IR-2 and thermotolerant strain EP-1. Hsie (1993) subjected thermotolerant yeast isolated from soil and sugarcane juice to cell fusion, giving fusants assignable to two groups: one growing rapidly above 40 °C but producing little ethanol, the other growing well around 39 °C with good ethanol yield. Laluce *et al.,* (1993) selected wild type yeast strains under pressures of temperature, high sugar and added ethanol and found that these were more thermotolerant than standard industrial strains and thermotolerance was affected by agitation rate.

2.2 Ethanol Tolerance

2.2.1 Effect of ethanol on cell growth and ethanol productivity:

Ismail and Ali (1971), studied the effect of ethanol concentrations in the medium upon yeast growth. At 12 per cent (v/v) ethanol all strains almost stopped growing. To obtain more tolerant strains the training method by successive transfers was used but was not successful. They also studied ethanol tolerance from genetic point of view. Rose (1980) reported that *Saccharomyces cerevisiae* could grow in ethanol concentrations varying from 8-12 per cent (v/v) ethanol. The efficiency of this organism to ferment glucose into ethanol up to a concentration of 12 per cent (v/v) was also highlighted. Gokhale *et al.,* (1983) isolated a *Saccharomyces* species which exhibited a specific ethanol productivity which was 70 per cent higher than that of *S. uvarum* ATCC 26602. Benitez *et al.,* (1988) examined strains of several *Saccharomyces* species for their ability to grow and ferment in a range of sucrose and ethanol concentrations. Alcoholic fermentation by yeasts is characterized by limitations due to ethanol and sugar inhibition. This phenomenon was studied for two yeasts *S. cerevisiae* and *S. bayanus* by Goma *et al.,* (1983).

The yeast growth and fermentation follow complex kinetics which result from the inhibition of growth rate and decline in cell viability. Inhibition kinetics for fermentation was less complex than that for growth and followed a classical non-competitive pattern. Beyond a certain critical concentration of ethanol cells do not grow and lose viability exponentially (Jones and Greenfield, 1984). Tyagi (1984) reported that trace amount of oxygen in the medium helps in increasing the viability of cells and productivity of ethanol. Otherwise as the alcohol concentration increases, cell viability decreases and hence productivity is lowered. Luong (1985) revealed that there were no striking differences between response of growth and ethanol fermentation and suggested that cells do not grow above 14 per cent (v/v) ethanol. A linear and non-linear Kinetic pattern was obtained for two strains of *S. cerevisiae* while studying the influence of ethanol concentration on the specific rate of ethanol production (Richter and Becker, 1985). Agudo (1985) reported that at least four genes are implicated in ethanol tolerance. The two characters, ethanol tolerance and ethanol production segregated independently but no ethanol-sensitive strains were able to produce high levels of ethanol. The effects of ethanol on the growth rates of two *Saccharomyces* strains was studied by Jones and Greenfield (1985). They reported that reduction in growth rate reflects a mixture of true inhibition and replicative inhibition. Gokhale *et al.,* (1986) compared two ethanol tolerant yeast isolates of *S. cerevisiae* for their invertase and alcohol dehydrogenase activities as well as ethanol productivity. The isolates showed significantly higher ethanol productivities compared to standard strains *S. uvarum* and other yeast strains tested. Specific and non-specific inhibitory effects of ethanol on yeast growth were studied by Jones and Greenfield (1987). Leite and Franca (1988) observed that the biomass and ethanol produced decreased with increasing alcohol concentration. Cells were freely permeable to ethanol and intracellular accumulation of ethanol was not detected. Richter (1989) reported that the loss of fermentative activity of yeast cells, observed in continuous fermentation experiments at increasing biomass concentration is explained by the assumption that the ethanol tolerance behaviour of microorganisms changes if a growth stabilizing factor limitation is present. Laplace *et al.,* (1991) studied combined alcoholic fermentation of D-xylose and D-glucose by strains of *Pichia stipitis, Candida shehatae, Saccharomyces cerevisiae* and *Zymomonas mobilis* with respect to sugar and produced ethanol tolerance. El-Diwany *et al.,* (1992) studied the effect of some fermentation parameters on ethanol production from beet molasses by *Saccharomyces cerevisiae* Y-7 and reported that yeast tolerated 10 per cent (v/v) ethanol but failed to grow at ethanol concentrations of 12 per cent or 15 per cent (v/v).

2.2.2 Intracellular and extracellular accumulation of ethanol:

Intracellular accumulation of 9.4 per cent (w/v) ethanol was more toxic to the cells than 13.8 per cent (w/v) ethanol added extracellularly to the fermentation broth (Nagodawithana and Steinkraus, 1976). Novack *et al.,* (1981) reported that intracellular accumulation of ethanol the in the cell against a concentration gradient causes a physiological stress in the cell. Intracellular ethanol is more detrimental to the functions such as growth and fermentation as compared to extracellular ethanol concentration. Jimenez and van Uden (1985) used the method of extracellular acidification for testing ethanol tolerance in yeasts. The effect of osmotic pressure on the intracellular accumulation of ethanol in *Saccharomyces cerevisiae* during wort fermentation was studied by D'Amore *et al.,* (1987). D'Amore *et al.,* (1988) also reported that nutrient limitation, and not necessarily intracellular ethanol accumulation, is a major factor responsible for the decreased growth and fermentation activities observed in yeast cells at higher osmotic pressures. The accumulation of ethanol in the cell and its exit from the cell was studied by Jones (1988).

2.2.3 Ethanol toxicity:

Girbes and Parrilla (1986) worked out a correlation between alcohol resistance and protein synthetic activity. They concluded that the sensitivity of the different yeast strains is determined primarily by the ability of the transport systems for the different nutrients to stand increasing ethanol concentrations. Jirku (1987) studied mechanism of ethanol toxicity, influence of osmotic pressure and temperature and the genetics of ethanol tolerance. Viel *et al.,* (1987) examined inhibitory behaviour of ethanol through systemic variations of glucose and alcohol concentrations. Their results show non-competitive inhibition of glucose metabolism for low ethanol concentrations (0-12% v/v). Bechard *et al.,* (1987) also investigated the toxicity effects of alcohols on *S. cerevisiae*. Anaerobic growth of *S. carlsbergensis* LAM 1068 was inhibited completely at the ethanol concentration of P_{mg} = 95 g/l (Toda *et al.,* 1987). Salgueiro *et al.,* (1988) reported that ethanol stimulated the leakage of amino acids and certain light absorbing compounds from the cells of *S.*

cerevisiae. Loureiro-Dias and ii a Santos (1990) monitored the effect of ethanol on intracellular pH and metabolism of a respiratory deficient mutant of *S. cerevisiae* by in vivo ^{31}P and ^{13}C NMR.

2.2.4 Formation/isolation of ethanol tolerant strains:

Seiko *et al*., (1985) isolated yeast strains having tolerance to inhibitive conditions such as low pH, high temperature, high ethanol and high sugar concentrations. Jimenez and Benitez (1988) formed hybrids between naturally occurring wine yeast strains and laboratory strains to increase genetic variability for improving the ethanol tolerance of yeast strains. Protoplasts of two NTG treated strains of *Pachysolen tannophilus* and *Saccharomyces* species were fused to get a fusant which produced 9.8 per cent (v/v) ethanol from molasses medium or 10.8 per cent (v/v) ethanol from raw sugar medium (Wang and Wang, 1989). Ernandes *et al*., (1990) isolated new ethanol-tolerant yeast strains from crude recycled yeasts for fuel ethanol production from sucrose. Kumar *et al*., (1990) studied the effect of ethanol on cell wall antigens of *S. cerevisiae*, and by making use of a specific test, isolated high ethanol producing strains. Argiriou *et al*., (1992) isolated two ethanol resistant strains of *S. cerevisiae* which could produce more than 16 per cent (v/v) ethanol using musts from raisin and sultana grapes. Laluce *et al*., (1993) selected wild type yeast strains under pressures of temperature, high sugar and added ethanol and studied their thermotolerance behaviour in sugar cane syrup fermentations.

2.2.5 Role of plasma membrane lipid composition in ethanol tolerance:

Plasma membrane is a biomembrane that acts as a function of osmosis and plays an important role in transport of metabolites to and from the environment within its vicinity and prevents the accumulation of toxic metabolites that may cause cell death (Hunter and Rose, 1971). Arthur and Watson (1976) reported that there is a direct correlation between the growth temperature and the degree of membrane lipid unsaturation, the lower the temperature, the greater the degree of lipid unsaturation. Mono-unsaturated fatty acids ($C_{16:1}$) was observed to be more effective in ethanol resistance than oleyl ($C_{18:1}$) or cetoleyl ($C_{20:1}$) residue. Linoleic acid ($C_{18:2}$) was even more effective due to increased degree of elasticity of fatty acids (Thomas et *al.*, 1978). Exogenous lipids supplement enhanced ethanol production as well as the endurability of cells. Yeast cells low in ergosterol but enriched with unsaturated fatty acid residues produced high concentration of ethanol (13-15.5% v/v) at substrate conversion efficiency nearly 90 per cent . Brown *et al.*, (1981) studied the relative rate of destruction of various cell wall structural components in presence of ethanol which showed the pattern: Phosphatidylcholine > phosphatidylethanol amine > phosphatidyl serine > phosphatidyl inositol, which is evident enough to prove the role of fatty acyl residues of plasma membrane as determinants of ethanol tolerance. Beaven *et al.*, (1982) worked on production and tolerance of ethanol in relation to phospholipid fatty acyl composition in *Saccharomyces cerevisiae* NCYC 431. They observed that during growth up to 32h, there was a progressive decrease in fatty acyl unsaturation in phospholipids, and a corresponding proportional increase in saturation.

Ohta and Hyashida (1983) reported that the addition of unsaturated fatty acids appears to be responsible for enhancing the growth and fermentation ability of sake yeast, whereas ergosterol confers ethanol tolerance. Some vegetable oils when supplemented in cane molasses stimulated utilization of sugars and increased alcohol production rate and yield, the effects were in general greater than with mixed fatty acids derived from oils (Saigal and Vishwanathan, 1984). Goyal (1985) reported that strains of *S. cerevisiae* enriched with palmitoleic acid were found to tolerate and ferment glucose to ethanol to give a final ethanol concentration of 12 per cent (v/v) during molasses fermentation in the presence of 6-7 per cent initial ethanol concentration at 30 °C. As lipids had been shown to protect yeast cells against ethanol toxicity, Walker-Capriogilo *et al.,* (1985) tested sterols, fatty acids, proteins and combinations of these, however, protection from growth inhibition was not seen. Freitas *et al.,* (1986) observed that the fermentative activity of a *S. cerevisiae* strain supplemented with oleic acid and ergosterol was less affected by different concentrations of ethanol than when supplemented with linoleic acid and ergosterol.

D' Amore and Stewart (1987) showed that plasma membrane phospholipids play an important role in ethanol tolerance mechanism. Increases in membrane unsaturated fatty acids result in increased yeast ethanol tolerance. Supplementation of growth media with combinations of unsaturated fatty acids, vitamins and proteins and physiological factors such as mode of substrate feeding, intracellular ethanol accumulations, temperature and osmotic pressure, all contribute to ethanol tolerance of yeast. Chadha (1987) reported that the incorporation of yeast cells with palmitoleic acid, Tween-80 and ergosterol resulted in higher ethanol yield.

Jimenez and Benitez (1987) gave a report on adaptation of yeast cell membrane to ethanol. The products of alcoholic fermentations, n-Butanol and iso-amyl alcohol, were found to act in a synergistic manner with each other and with ethanol in causing cell death in suspensions of non-growing cultures of *S. cerevisiae* (Okolo *et al.*, 1987). Ghareib *et al.*, (1988) described the relationship between ethanol tolerance of *S. cerevisiae* and its lipid content and composition. Mishra and Prasad (1989) reported that *S. cerevisiae* became more resistant to ethanol with an increase in unsaturated fatty acids and that membrane fluidity could be an important determinant of ethanol tolerance. Viegas *et al.*, (1989) evaluated the inhibition of growth of yeast by octanoic or decanoic acids in *S. cerevisiae* and *Kluyveromyces marxianus* in association with ethanol.

Novotny *et al.*, (1992) found that raising cellular sterol levels by genetic and physiological means resulted in an increase in ethanol tolerance in a yeast population. On the other hand, Agudo (1992) suggested a relationship between membrane fluidity and ethanol tolerance but did not support a direct role of unsaturated fatty acids in this tolerance.

2.3 Genetic Improvement Of Yeast

The subject of yeast genetics is currently being applied to industry in order to (1) identify genes that control the behaviour of yeast strains, (2) elucidate the genetic control of complex metabolic processes such as the uptake and utilization of different substrates, and (3) produce new and improved yeast strains via breeding programs (Stewart, 1981a).

Stewart (1981b) reviewed the various techniques of genetic manipulation of industrial yeast strains. The author has compared hybridization, spheroplast fusion and transformation. Hybridization is a secondary technique for verifying the gene composition of recombinants produced by fusion and recombination. Fusion is more empirical than manipulative since little control can be exerted over the genotypic make-up of recombinant. Transformation with native DNA is a more subtle technique than fusion since the injection of single genetic trait into strains can be controlled.

Beach and Nurse (1981) have reported high-frequency transformation of a leu 1^- strain of the fission yeast *Schizosaccharomyces pombe* with hybrid plasmids containing the *Saccharomyces cerevisiae* LEU 2^+ gene. Heslot (1983) reported that the yeast species *S. cerevisiae*, *Schizosaccharomyces pombe* and *Kluyveromyces lactis* are amenable to genetic engineering. The former has been transformed by plasmids, able either to integrate into the chromosome of the recipient strain, or to replicate autonomously. Bardiya *et al.*, (1983) investigated yeast breeding programmes for higher alcohol production. They found that some of the spore clones produced higher alcohol than the parents.

Continuous fermentation of cheese whey with a catabolite repression-resistant mutant of *S. cerevisiae* was studied by Terell *et al.*, (1984). Morris and Roth (1985) have outlined some of the current techniques used for gene cloning in *S. cerevisiae*, describing transformation procedures, and the types of DNA cloning vectors either already or imminently in use. Industrial yeast strains are polyploid/aneuploid and consequently are less readily modified by classical genetic techniques such as hybridization. Thus for the genetic improvement of polyploid yeasts, techniques such as rare mating, spheroplast fusion and DNA transformation can be employed (Panchal *et al.*, 1986). Ribeiro *et al.*, (1989) constructed new yeast strains carrying the genetic information for different amyloytic enzymes to simplify ethanol production from starchy substrates.

Polsinelli (1990) has discussed genetic improvement of wine yeast *S. cerevisiae* with reference to mutation, selection, spheroplast fusion, cytoduction, transformation, application of genetic engineering techniques and possibilities for production of new yeast strains of value in wine-making. A recombinant strain of *S. cerevisiae* (strain RD) was obtained through conjugation of one ethanol-tolerant strain isolated from grape juice, and one strain that produces glucoamylase. Substrate consumption, ethanol productivity and ethanol yield were determined in fed batch/batch culture (Krauel *et al.,* 1990). Neelam and Amarjit (1991) studied ethanol production by thermotolerant yeast and its UV resistant mutants. Takuma *et al.,* (1991) described isolation of a xylose reductase gene from *Pichia stipitis* and its expression in *S. cerevisiae*. However, the transformed *S. cerevisiae* was not capable of ethanol production from xylose.

A *S. cerevisiae* strain was transformed with mechanically fragmented *Kluyveromyces marxianus* DNA using intact cells or protoplast transformation methods. The transformed strain was able to assimilate lactose (Abdel-Salam *et al.,* 1992). Porro *et al.,* (1992) reported 95 per cent utilization of lactose by transformed *S. cerevisiae* cells. Genetic manipulation of *Zymomonas mobilis*, aimed at extending the range of possible substrates and thus improving its industrial utility was reported by Yanase (1993).

2.4 Yeast Protoplast Studies

The potential use of fungal protoplasts as an experimental system has been exploited over the past (Perberdy, 1979). Protoplast fusion has proved to be potentially valuable in the development of improved strains of

Saccharomyces for brewing, distilleries, wine making, baking and production of food and fodder yeasts. The technique is universally applicable and has been used for a wide range of bacteria, fungi, yeasts and algae. It provides a means of genetic recombination in micro-organisms in which a natural system of gene transfer has not been demonstrated. Thus among the wide variety of strain improvement techniques, protoplast fusion clearly has a great potential. It will not be surprising if protoplast fusion between distantly related strains is used to generate modified yeast . strains capable of converting complex substrates like cellobiose, starch, etc. to either ethanol or single cell protein rich in required amino acids (Sathe *et al.*, 1992; Kumari and Panda, 1992).

Giaja (1919) first reported protoplasts from yeast cells using gastric juice derived from Helix pomatia. Since then many studies have been carried out with respect to protoplast fusion in yeast as evidenced by reviews, articles and papers published in the last few years (van Solingen and Van der Plaat, 1972; Figueroa *et al.*, 1984; Russel *et al.*, 1986; Farahnak *et al.*, 1986; Sakai *et al.*, 1986; Tamaki, 1986; Hoffman *et al.*, 1987; Chung *et al.*, 1987; Lee *et al.*, 1980; Janderova *et al.*, 1990).

Commercially available enzyme preparations are generally used for isolation of protoplasts from yeasts. The enzymes such a glusulase, sulfatase and helicase are prepared from digestive juice of the snail, *Helix pomatia*. Another enzyme Zymolyase is derived from *Arthobacter luteus* (Sathe *et al.*, 1992). The enzyme mutanase is an α-1, 3-glucan glucanohydrolase and is derived from a strain of *Trichoderma harzianum*. It contains cellulase, xylanase, β-1, 3-glucan glucanohydrolase and neutral protease activities (Hamlyn *et al.*, 1981; Stephen and Nasim, 1981; Dickinson and Isenberg,

1982). This enzyme is used under the commercial name Novozym 234. Culture age affects protoplast release in *Saccharomyces cerevisiae.* Stationary phase cells are resistant to lysis while exponential phase cultures are readily converted to protoplasts (Deutch and Parry, 1974). Brown (1971) compared the susceptibility of the cells of *Torulopsis glabrata, Candida albicans* and *S. cerevisiae* to the action of helicase. The susceptibility diminished gradually during the transition from exponential to stationary phase of growth. Shahin (1971) reported a high percentage of protoplasts from stationary phase cells of five strains of *Schizosaccharomyces pombe* using the commercial preparation helicase obtained from *Helix pomatia.* Shahin (1972) also reported that a number of generations must occur in the presence of glucose before daughter cells of stationary or old cultures of *Saccharomyces* became amenable to protoplast formation by the action of snail enzyme.

Kopecka (1975) formed protoplast of *Schizosaccharomyces pombe* and *Schizosaccharomyces versatilis* by the combined application of snail enzymes and *Trichoderma viride* enzyme. About 100 per cent protoplast population was obtained in 30 min. The protoplasts were viable and could regenerate cell walls and revert into normal cells. A new method for the efficient conversion of *Sc. pombe* cells into protoplast was developed to study their guanine uptake (Housset *et al.,* 1975). Ferenczy *et al.,* (1976) have studied the factors influencing the fusion frequency of protoplast of auxotrophic mutants of *Aspergillus niger.* Van Solingen and Van der Plaat (1977) fused spheroplasts of two different auxotrophic strains of *S. cerevisiae,* both of mating type a. After regeneration the resulting zygotes were found to be diploid cells of mating type aa. Yamamoto and Fukui (1977) also presented a technique for the fusion of protoplasts prepared from the same mating type cells of two auxotrophs of *S. cerevisiae*. Sipiczki and Ferenczy (1977) have suggested from the results of haploidization induced with parafluorophenylalanine or of random spore analysis that the fusion products of protoplasts prepared from haploid cells of *Sc. pombe* were diploids.

Schwencke and Nagy (1978) reported that snail enzyme preparation failed to transform cells of *Schizosaccharomyces* into protoplast even under a wide range of conditions. This failure is related to the particular composition of the cell walls of this yeast. The fusion of protoplasts prepared from two different haploid strains of mating type a, carrying different nuclear and mitochondrial genetic markers was done by Gunge and Tamaru (1978). They have described a detailed genetic analysis of fusion products of protoplasts in *S. cerevisiae*.

Fusion products having mixed parental characteristics have been reported when protoplast of *S. cerevisiae* and *Saccharomycopsis lipolytica* were fused (de van Broock *et al.*, 1980). Stephen and Nasim (1981) reported that mutanase (Mutanase Novo) affects the high frequency production of protoplasts in the yeasts *S. cerevisiae, Sc. pombe, Kluyveromyces lactis, Trichosporon pullulans* and *Schwanniomyces alluvius.* Regeneration frequencies varied with the strain used and ranged between 10 and 18 per cent. Hamlyn *et al.,* (1981) compared several commercial polysaccharases for their ability to liberate protoplasts from fungi. Dickinson and Isenberg (1982) described a method for conversion of *Sc. pombe* cells to spheroplast using Novozym 234. Wilson and Ingledew (1982) made an attempt for intergenic protoplast fusion between *Schwanniomyces alluvius* and *S. cerevisiae* to obtain hybrids with both amylolytic activity and improved fermentation ability. Hybrids formed were, however, unstable and spontaneously segregated into original auxotrophic parent strains.

Figueroa *et al.*, (1984) selected *S. cerevisiae* and *S. diastaticus* for protoplast fusion. *S. diastaticus* carried spontaneous petite mutation and could not metabolize starch unlike respiratory competent *S. diastaticus* from which it was derived. Respiratory competent fusion products obtained were capable of using dextrins and starch as carbon source and were able to produce 3.79 per cent ethanol in medium containing starch as compared to 2.30 per cent ethanol produced by parent *S. diastaticus*. Later, they again tested fermentation of starch based industrial media with yeast protoplast fusion products (Figueroa *et al.,* 1985). To improve starch fermenting ability of *S. diastaticus*, Sakai *et al.*, (1986) used protoplast fusion technique and found that some fusants produced more ethanol and amylase than the parental strains. Tamaki (1986) carried out protoplast fusion between various auxotrophic haploid strains of fast fermenting *S. cerevisiae* and starch fermenting *Schwanniomyces castellii.* The fusion hybrids showed starch fermentability which was derived from one of the fusion partners, *Sch. castellii.* The fusion hybrids also showed excellent sporulation and spore germination.

Pina *et al.*, (1986) obtained intergenic hybrids of *S. cerevisiae* and *Zygosaccharomyces fermentati* by protoplast fusion. The hybrids presented characteristics of both parents and could have important industrial applications as good fermenting strains. The protoplast fusion technique was used to construct hybrids between auxotrophic strains of *S. cerevisiae* having high ethanol tolerance and an auxotrophic strain of lactose-fermenting *Kluyveromyces fragilis* (Farahnak *et al.,* 1986). The fusants obtained were prototrophic and capable of assimilating lactose and producing ethanol in excess of 13 per cent (v/v).

Polyethylene glycol induced protoplast fusion of two haploid strains of *S. cerevisiae* and *S. diastaticus* was carried out by Hoffmann *et al.,* (1987). Laluce *et al.,* (1989) made an effort to improve the thermotolerance of amylolytic yeasts by protoplast fusion. Protoplast fusion between sake and wine yeasts was done for the development of young white wine with novel flavour characteristics (Yokomori *et al.,* 1989). Ribeiro *et al.,* (1989) fused protoplasts of two auxotrophic derivatives of 4D strain of *Saccharomyces* which showed improved performance of ethanol production from starch. The possibility of utilizing starch splitting yeast on an industrial scale for the production of biomass lipids has also been discussed by Janderova *et al.,* (1989), using protoplast fusion and genetic engineering of the production strain. Wang and Wang (1989) reported improvement of ethanol tolerance of xylose-fermenting yeast by protoplast fusion.

Menzel *et al.,* (1990) studied the effect of an enzyme preparation of *Streptomyces* having cell wall lytic activity on cells of *Saccharomyces cerevisiae* and found that only protoplasts and no spheroplasts were formed. Spencer *et al.,* (1990) constructed industrial yeast strains by protoplast fusion using *S. diastaticus* for starch utilization and *S. kluyveri* for melibiose utilization. Spencer *et al.,* (1990) also reviewed protoplast fusion techniques to obtain yeast hybrids with improved properties, using fusogenic agents such as polyethylene glycol. They produced a *S. cerevisiae* strain that ferments lactose to ethanol by transferring β-galactosidase and lactose permease in successive fusions from *Kluyveromyces lactis*. Svoboda and Ourednicek (1990) have described a method of yeast protoplast immobilization and subsequent isolation from calcium alginate gel. The regeneration of the protoplast cell wall and concomitant reversion to normal growth and morphology is a pre-requisite for fusion and transformation studies. Factors affecting the regeneration of such Novozym 234-prepared protoplasts have been discussed by O'Brien and Whittaker (1990). The possibility of increasing ethanol productivity by constructing intraspecific and interspecific hybrids by the fusion of protoplasts or spheroplasts was investigated by Kavanagh and Whittaker (1990). They also studied the effect of buffer viscosity in the rate of protoplast formation in the yeasts *Pachysolen tannophilus* and *S. cerevisiae* (Kavanagh and Whittaker, 1991).

Wang *et al.,* (1992) selected D-xylose and cellobiose-fermenting and ethanol-producing strains by electric field-induced protoplast fusion between strains of *Candida guilliermondii* and *S. cerevisiae*. A thermotolerant flocculating yeast, *S. cerevisiae* KF-7 was constructed by protoplast fusion of the flocculating yeast *S. cerevisiae* IR-2 and the thermotolerant yeast *S. cerevisiae* EP-1 (Kida *et al.,* 1992). Grover (1992) carried out protoplast fusion between *Endomycopsis fibuligera* and *S. cerevisiae* to derive a yeast strain which could convert starchy substance directly to ethanol. Sathe *et al.,* (1992) reviewed protoplast fusion as a technique for strain improvement in the yeast *S. cerevisiae*. Dilbaghi (1994) fused protoplasts of *S. cerevisiae* and *S. diastaticus* for efficient utilization of starch for ethanol production.

2.5 Ethanol Production

2.5.1 Substrates:

The production of alcohol from food processing wastes has drawn considerable interest due to the rising energy cost and the negative cost values of wastes as substrates (Hang *et al.,* 1981). The impact of various raw materials and alternate fermentation methods on process designs for the production of fuel grade ethanol have been discussed by Prouty *et al.*, (1980). The wastes from sugar cane industry, dairy industry and food processing industry have been successfully utilized for ethanol production.

Acid saccharification and alcohol fermentation of unripe banana fruit has been reported by Pontiveros *et al.*, (1978). Tewari *et al.*, (1982) have produced ethanol from waste potato. Sugar beet can be used as an alternate or as a supplementary raw material in bioethanol generation (Goyal and Tauro, 1983). They reported production of 8.6 per cent ethanol within 24 hours from beet juice using a selected yeast strain. The effect of nine different synthetic culture media was investigated by Abdel-Fattah *et al.*, (1984) for ethanol production by *S. cerevisiae*. Banana peels saccharified with sulfuric acid, cellulase and steam were fermented with *S. cerevisiae* for ethanol production (Tewari *et al.*, 1985; Tewari *et al.*, 1986).

Menezes *et al.*, (1985) utilized bamboo stems treated with commercial amylolytic enzyme and a cellulolytic fungi for fermentation using *S. cerevisiae*. Whalen *et al.*, (1985) described a process for co-fermentation of whey and corn to produce industrial alcohol. Fermentation of whey to ethanol using "sake" brewing yeasts was reported by Suzuki *et al.*, (1985). Abouzeid and Reddy (1986) investigated direct fermentation of unhydrolyzed potato starch to ethanol by *Aspergillus niger* and a mixed culture of *A. niger* and *S. cerevisiae*. Amylolytic activity, rate and amount of starch utilization and ethanol yields increased several-fold in co-culture versus the monoculture due to the synergistic metabolic interactions between the species. Grape pomace was used as a substrate for the production of ethanol under solid state fermentation conditions (Hang *et al.*, 1986). The yield of ethanol amounted to greater than 80 per cent of the theoretical based on the fermentable sugar consumed.

Moriya *et al.,* (1987) isolated yeasts and then investigated the conditions of ethanol fermentation of beet molasses using two strains of *S. cerevisiae*. They observed that agitation of mash and stepwise addition of beet molasses accelerated the ethanol production and fermentation yield. Dallmann *et al.,* (1987) have reported continuous fermentation of apple juice by immobilized yeast cells. Fermentation of fruits and cladodes of prickly pear cactus for ethanol production was reported by Retamal *et al.,* (1987). Bhatt *et al.,* (1987) compared *Candida krusei* and two strains of *S. cerevisiae* for ethanol production from mixed fruit juice of damaged guava and banana. Whalen (1988) described a continuous fermentation process its development and scale-up for the co-fermentation of cheese whey and corn for ethanol production. Akpan *et al.,* (1988) have reported production of ethanol from cassava whey. Yeasts, isolated from fruits, vegetables, flowers and sugarcane juice, were screened for the production of ethanol from potato mash and for their efficiency to utilize lactose (Tewari *et al.,* 1988). Nain and Rana (1988) carried out tests to find the optimum operating conditions for production of ethanol from sugar beet juice by *S. cerevisiae*. A review on ethanolic fermentation of whey and whey molasses mixtures was presented by Zakrzewski and Zmarlicki (1988).

Gibbons and Westby (1989) reported co-fermentation of sweet sorghum juice and grain for production of fuel ethanol and distillers' wet grain. The use of substandard fruits and vegetables for ethanol production has been reported by Cheema (1989), Nandita (1991), Kaur (1992) and Singh (1993). *Tewari et al.,* (1989) reported comparative saccharification of vegetable wastes from cauliflower stalks and pea-peels by acid, enzyme and steam and subsequent fermentation of the hydrolyzed mashes for ethanol production by *S. cerevisiae*. Patil *et al.,* (1989) reported that novel supplements like skim milk, chitin and fungal mycelium, enhance the ethanol production in cane molasses fermentation by recycling yeast cells. Ameh *et al.,* (1990) isolated yeast strains from fermented beverages, foods and fruits and tested their fermentative activity. A strain of *S. cerevisiae* was specifically designed for efficient commercial production of ethanol from beet molasses (Saito *et al.,* 1990). Diez-Jerez *et al.,* (1990) tested sugarbeet media with different sources of nitrogen and phosphorous for ethanol production by *Zymomonas mobilis* and *Saccharomyces uvarum*. Park and Baratti (1991) performed similar experiments using *Zymomonas mobilis* alone. Hariantono *et al.,* (1991) investigated alcoholic fermentation from raw maize starch using a strain of *Schizosaccharomyces pombe* and a raw starch saccharifying enzyme from *Corticium rolfsii.*

2.5.2 Organisms used for fermentation:

Although many bacteria and fungi have been tested for ethanol production from different substrates, *Saccharomyces cerevisiae* is the only organism currently being used for industrial ethanol production.

Fernandez *et al.,* (1981) reported alcohol production by eight yeasts, *Kluyveromyces fragilis, K. marxianus, Saccharomyces bayanus, S. carlsbergensis, Candida pseudotropicalis* and three strains of *S. cerevisiae*. The fermentation of sucrose based raw materials has been studied using immobilized cells of *Zymomonas mobilis* in a lab scale bioreactor (Grote and Rogers, 1985). Rao and Mutharasan (1986) studied alcohol production by *Clostridium acetobutylicum* induced by methyl viologen. Drawert and

Schwank (1986) tested seven yeast strains for fermentation of lactose in a glucose lactose medium. Significant growth of the yeast species *Kloeckera apiculata, Candida stellata, C. colliculosa* and *C. pulcherrima* was noted during fermentation of some Australian wines carried out by inoculated and indigenous yeasts (Heard and Fleet, 1986). Direct conversion of cellulose to ethanol using *Neurospora crassa* was reported by Deshpande *et al.,* (1986). *Candida krusei* and two strains of *S. cerevisiae* were compared for their production from juice of damaged fruits (Bhatt *et al.,* 1987).

Baranaik (1988) screened type strains of 24 yeast species for fermentation of D-xylose. In most strains tested ethanol production was negligible. A *Zymomonas mobilis* strain was compared with a *Saccharomyces* sp. for the fermentation of sugarcane molasses. Both gave similar final ethanol concentrations but fermentation was slower with *Z. mobilis* than with *Saccharomyces* sp. (Murphy *et al.,* 1988). Shamala and Sreekantiah (1988) described the use of wheat bran as a nutritive supplement for the production of ethanol by *Z. mobilis.* Raghav *et al.,* (1989) compared kinetics of fermentation by standard *Saccharomyces uvarum* strain, an isolate of *S. cerevisiae* and a highly flocculant strain of *S. uvarum* in batch mode. Results showed that both the isolate and the flocculant strain had more desirable characteristics than the standard strain for ethanol production from cane molasses. Millichip (1989) studied ethanol production by *Zymomonas* cultures in yeast-conditioned media. Agrawal and Veeramallu (1990) reported ethanolic fermentation by *Z. mobilis* ATCC 10988 in repeated batch cultures. Gu (1992) tested *Z. mobilis* for industrial alcohol production. On starch, hydrolysate fermentation rate and ethanol yield exceeded those obtained using *Saccharomyces* sp. Molasses could also be used, but pretreatment with activated carbon was necessary to remove substances unfavourable to fermentation.

2.5.3 Factors Influencing fermentation:

The fermentation of various substrates to ethanol is dependent on a large number of environmental factors like substrate concentration, pH, temperature, oxygen, medium composition, inoculum concentration and also on the mode of fermentation used viz. batch or continuous mode with or without cell recycle. The effect of such factors on fermentation have been reviewed.

Verma *et al.,* (1983a.b) used cell recycle technique for ethanol production from five strains of *S. cerevisiae*. A 10-80 per cent reduction in fermentation time was observed thus increasing the overall alcohol production per unit time. In addition, 5.7-6.8 per cent sugars could be saved which otherwise were wasted for raising inoculum. Modi and Shankar (1983) have reported continuous ethanol production using *S. cerevisiae*. They observed that at higher dilution rates efficiency was higher with decreased sugar utilization. Karanth (1983) reviewed some high productivity processes considered industrially feasible and their important features. In a low dilution rate study, Parsons *et al.,* (1984) suggested a low optimum pH for ethanol fermentation by yeast. Maiorella *et al.,* (1984) reported that inhibition by secondary feed components can limit productivity and restrict process options for the production of ethanol by fermentation. The effect of trace amounts of oxygen on the degree of ethanol inhibition in a continuous anaerobic culture of *S. cerevisiae* was studied by Hoppe and Hansford (1984). Cheryan and Mehaia (1984) reported that ethanol productivity was greatly enhanced by using a membrane bioreactor which was configured as a cell recycle system.

Kaeppeli and Sonnleitner (1986) reported the regulation of sugar metabolism in *Saccharomyces* type of yeast. Continuous culture experiments revealed that oxidative glucose metabolism is possible at low growth rates. A flocculant *Saccharomyces* sp., isolated from spontaneously fermenting sugarcane juice, was used to produce ethanol in a continuous upflow reactor (Abate *et al.,* 1987). Dorsemagen *et al.,* (1988) have described the first production-scale Hoechst process for continuous ethanol production which began operation in Brazil in 1986. Zikmanis *et al.,* (1988) observed that dehydrated and rehydrated cells of *S. cerevisiae* showed significantly higher ethanol production from exogenous substrate than intact yeast cells under both anaerobic and aerobic conditions. Weide *et al.,* (1989) reported that the addition of a biolipid extract to the fermentation medium increased the yield of ethanol.

Experiments conducted by Vajpeyi *et al.,* (1990) showed that yeast can be recycled for a prolonged period without adversely affecting its activity, and that the cycle time can be shortened to eight hours. Patil and Patil (1990) reported that adding fungal mycelium from various species or waste mycelium from antibiotics industry accelerated ethanol production from sugarcane molasses by *S. cerevisiae*. Ethanol production from cane molasses medium was also reported by Chen (1990), Kida *et al.,* (1991) and Yusuf *et al.,* (1993). Melzoch *et al.,* (1991) carried out a series of continuous experiments with *S. cerevisiae* in a membrane bioreactor. New yeast strains for ethanol production from molasses at higher sugar concentration were screened by Bertolini *et al.,* (1991) and Szopa *et al.,* (1993).

2.6 Production of Wine

The fermentation of fruit juices by yeast results in production of wine (Amerine and Singleton, 1968). It contains vitamins, amino acids, esters, sugars, tartaric acid, etc. Naturally fermented wines usually contain about 8 to 14 per cent alcohol. Wine has deeply penetrated the social fabric and culture of times and countries from which we spring. Somras (wine) is our Vedic drink (Tewari, 1980).

Alian and Musenge (1970) and Tewari *et al.,* (1987) have reported wine production from pineapple waste. Tewari *et al.,* (1978) evaluated the enological properties of grapes grown in Punjab state for the production of wine. The feasibility of processing of grapes in Punjab state and the establishment of experimental winery, Punjab Marketing Federation, Chandigarh, has been reported by Tewari (1978). There has also been a report on the establishment of pilot scale grape processing plant for wines and brandy in Punjab (Tewari, 1978). Wines and vinegars have been used as food and medicine since time immemorial (Tewari and Gupta, 1978). A manual for diabetics has been published by Tewari (1979c). Wine from *Momordica charantia* developed by Tewari (1979c, 1980 and 1986) also has medicinal value for diabetics.

Vyas and Joshi (1982) and Tewari *et al.,* (1988) described a method for making an acceptable quality of plum wine. The influence of fruit composition, maturity and mold contamination on the colour and appearance of strawberry wine have been discussed by Pilando *et al.,* (1985). Minarik (1985) reported some microbiological and biotechnological problems in wine making. Heard and Fleet (1986) examined the occurrence and growth of yeast species, that occur naturally in grape juice, during fermentation of some Australian wines. Wine from banana has been chemically and seasonally analyzed by Stein (1986). The utilization of substandard pears for the production of wines as food and medicine was reported by Tewari *et al.*, (1987). *Saccharomyces* and *Schizosaccharomyces* species isolated from palm wine were used by Obisanya *et al.*, (1987) for producing wine from mango. The production of bottle-fermented sparkling wine using yeasts immobilized into calcium alginate beads was reported by Fumi *et al.*, (1988). Reed and Nagodawithana (1988) have described the technology of yeast usage in wine making.

Yokomori *et al.*, (1989) suggested that significant difference in type and quality of wine was brought about by wine yeast used for fermentation of grape must. They bred hybrid strains of sake and wine yeasts for development of young white wine with novel flavour characteristics. Cheema (1989) has used waste guava for the production of quality wine. A process for thermally controlling wine-making was described by Laude-Bousqet (1989). Lembke *et al.*, (1990) gave a process for sparkling wine production. The production of wine from grapes could be improved by the addition of honey and sulfur dioxide in a total amount effective to inhibit browning discolouration (Lee and Kime, 1990). The application of active dry yeast in the technology of sparkling wines was described by Nemecek *et al.*, (1990). Ozilgen *et al.*, (1991) studied the kinetics of spontaneous wine production. Nandita (1991) studied wine and fuel alcohol production from waste mangoes. Argiriou *et al.*, (1992) isolated alcohol resistant strains of *S. cerevisiae* for potable alcohol production using molasses. The production of wine from ginger has been reported by Singh (1993).

Chapter

3. MATERIALS AND METHODS

3.1 Organisms

The yeast strains used in the present studies on "Improvements in efficiency of yeasts for ethanol production" were collected with courtesy from the sources given below.

Table 3.1 Yeast Strain Source

Sr. No.	Yeast	Strain No.	Courtesy
1	*Saccharomyces cerevisiae*	220	Deptt. of Microbiology, Punjab Agricultural University, Ludhiana
2	*S. cerevisiae*	360	
3	*S. cerevisiae*	MTCC 181/3c	Deptt. of Biotechnology, Punjabi University, Patiala
4	*Trichosporon beigelli*	BR-1	
5	*Candida haemulonii*	2(3)	
6	*C. parapsilosis*	6(2)	
7	*Candida sp.*	3(1)	
8	*Candida sp.*	5(2)	

3.1.1 Maintenance of Cultures

All the yeast cultures were maintained on Glucose Yeast Extract (GYE) medium (Tuite, 1969), having the following composition.

Table 3.2 GYE Composition

Glucose	10.0 g
Yeast Extract	5.0 g
Peptone	5.0 g
Agar	20.0 g
Distilled water	1000 ml
pH	5.0

These cultures were stored in a refrigerator and subculturing was done after every 15-20 days.

3.2 Screening of Yeast strains for Ethanol production

Cane sugar was used as the substrate for screening yeast strains no 1-8 for their efficiency to utilize substrate for the production of ethanol and ginger wine at elevated fermentation temperatures.

3.2.1 Preparation of Inoculum

Test tubes, each containing 10 ml of glucose yeast extract broth, were inoculated with loopful of actively growing (24 h) cultures of *S. cerevisiae-220, S. cerevisiae-360, S. cerevisiae MTCC-181/3c, Trichosporon beigelli, Candida haemulonii, C. parapsilosis, Candida sp.-3(1)* and *Candida sp.-5(2)*. The tubes were subsequently incubated at 30 ± 1 °C for 24 h. Inoculum (5 ml), containing yeast was transferred to 100 ml of glucose yeast extract broth in 250 ml capacity Erlenmeyer flasks. The flasks were then incubated at 30 ± 1 °C for 24 h and the contents were used as inoculum for screening of the strains of yeast for ethanol fermentation.

3.2.2 Fermentation of Ethanol

Fermentation of ethanol was carried out in 250 ml capacity Erlenmeyer flasks, each containing 100 ml of fermentation medium adjusted to pH 5.0. Each flask was inoculated at the rate of 15.0 % inoculum and fermentation was carried out at different temperatures, viz. 30 ± 1 °C, 38 ± 1 °C and room temperature 37 to 42 °C. The utilization of the substrate was recorded and expressed as total soluble solids at regular intervals till constant reading was obtained. All the experiments have been conducted in duplicate.

3.3 Temperature Tolerance Profile Studies on Yeast

The growth of *S. cerevisiae-220* and of *S. cerevisiae-360* has been studied at 30 ± 1 °C and 38 ± 1 °C. Optical density of the cultures has been recorded using a spectrophotometer (Bausch and Lomb Spectronic – 20).

3.3.1 Primary Inoculum

A loopful of actively growing culture (24 h) of *S. cerevisiae-220* and *S. cerevisiae-360* was transferred into tubes, each containing 10 ml of glucose yeast extract broth. The tubes were incubated at 30 ± 1 °C and 38 ± 1 °C for

20 h. Yeast inoculum (2 ml) was transferred from each of the tubes into 50 ml of glucose yeast extract broth in 250 ml capacity conical flasks. The flasks were incubated at 30 ± 1 °C and 38 ± 1 °C for 20 h. The contents of these flasks have been used as inoculum for studying the effect of temperature on the growth of *S. cerevisiae-220* and *S. cerevisiae-360.*

3.3.2 Studies on growth of Yeasts

Erlenmeyer flasks (250 ml capacity) containing 100 ml of glucose yeast extract broth were inoculated with 2 ml of the primary inoculum and have been incubated at 30 ± 1 °C and 38 ± 1 °C separately, in duplicate.

The growth was recorded in terms of optical density at 540 nm using a spectrophotometer (Bausch and Lomb Spectronic – 20) at regular intervals of 2 h for a period of 32 h followed by a control till constant optical density readings were achieved.

3.4 Ethanol Tolerance Profile Studies on Yeast

The effect of initial ethanol concentration on growth of *S. cerevisiae-220* was studied by supplementing glucose yeast extract broth with ethanol at a concentration of 10.0, 11.0, 12.0, 13.0, 14.0 and 15.0% (v/v), whereas a concentration of 8.0, 10.0, 12.0, 13.0, 14.0 and 15.0% (v/v) has been initially maintained in the medium for *S. cerevisiae-360*, Inoculum (2 ml) prepared for respective strains was aseptically transferred to 150 ml of glucose yeast extract broth supplemented with different ethanol concentrations in 250 ml capacity Erlenmeyer flasks. All the flasks were incubated at 30 ± 1 °C.

The growth was recorded in terms of optical density at 540 nm using a spectrophotometer (Bausch and Lomb Spectronic – 20) followed by a control. The observations have been recorded at every two hour intervals, up to 46 h of incubation. All the experiments were performed in duplicate equipment.

3.5 Studies on Protoplast Fusion

3.5.1 Preparation of Protoplasts

Protoplasts from *S. cerevisiae-220* and *S. cerevisiae-360* were prepared according to the method of Dickinson and Isenberg (1982). The protocol for the preparation of protoplasts is given in Figure 1.

Cells of *S. cerevisiae-220* and *S. cerevisiae-360* were grown in 50 ml batches of glucose yeast extract broth at 30 ± 1 °C and 38 ± 1 °C respectively, for 20 h to a density of 5 x 10^7 cells ml^{-1}.

↓

Cells were harvested by centrifugation at 2000 rpm for 5 min and washed with distilled water.

↓

Cells were suspended separately in 100 nM-ETDA, 100 mM-2-(N-morpholino) ethane sulphonic acid (MES), pH 6.0, to which Dithiothreitol (DTT) had been added, just before use, to a final concentration of 5 mM and suspensions were incubated at 30 ± 1 °C with occasional shaking, for 20 min.

↓

Cells were harvested by centrifugation at 5000 rpm for 1 min and washed once with 0.6 M sorbitol, 20 mM-MES, pH 6.0.

↓

Cells were resuspended in protoplasting buffer containing 1 mg ml^{-1} Novozym 234, 0.6 M KCL, 0.01 M phosphate buffer, pH 5.85, to which Dithiothreitol (DTT) had been added, just before use, to a final concentration of 5 mM. The suspensions were incubated at 30 ± 1 °C with occasional shaking, for 60 min.

↓

After digestion, samples were taken, diluted 100-fold with 0.6 M sorbitol, 0.2 M KCL, 20 mM-MES, pH 5.0 (SKM buffer) and counted microscopically with the help of haemocytometer.

↓

Collected the protoplasts by centrifugation at 2000 rpm for 10 min.

↓

Stored the protoplasts in SKM buffer at 4 °C.

Figure 1 Protocol for preparation of Protoplasts

3.5.2 Protoplast Fusion and Regeneration

Fusion of protoplasts obtained from *S. cerevisiae-220* and *S. cerevisiae-360* was done according to the method of Gunge and Tamaru (1978). Fusion was induced with polyethylene glycol (PEG 6000) in the presence of calcium ions. Equal amounts of protoplast suspensions were mixed and centrifuged at 3000 rpm for 2 minutes. Polyethylene glycol (34%) containing 0.8 M sorbitol and 10 mM $CaCl_2$ was added to the pellet obtained. Incubation was done at 30 ± 1 °C for 50 min. The aggregated protoplasts were then spread on plates containing glucose yeast extract medium supplemented with 0.8 M sorbitol. Incubation was carried out at 38 ± 1 °C for 48 h.

3.6 Analysis

3.6.1 Total Soluble Solids (TSS)

The total soluble solids (°Brix) of juice and wine were determined by using Erma hand refractometer.

3.6.2 pH

The pH of juice and wine were determined by pH meter (Elico). The pH meter was calibrated with standard buffer solutions of pH 4.0 and 7.0 before use.

3.6.3 Total Acidity (Amerine et al., 1980)

Reagents:

a) Sodium hydroxide (0.1 N) - 4.0 g of sodium hydroxide pellets were dissolved in 1000 ml of distilled water.
b) Phenolphthalein indicator (1%) – 1.0 g of phenolphthalein was dissolved in 100 ml of ethanol.

Procedure:

Pipetted 5 ml of sample in a conical flask, diluted it to 50 ml using distilled water. Then added 2-3 drops of phenolphthalein indicator. Titrated it with 0.1 N NaOH, adding it dropwise from burette till one drop gives the end point i.e., from colorless to light pink. Recorded the amount of alkali used.

Calculations:

1 ml of 0.1 N NaOH = 0.015 g tartaric acid

$$\text{Total acidity (\% w/v)} = \frac{\text{Vol of 0.1 N NaOH used x 0.015 x 100}}{\text{Vol of sample taken}}$$

Equation 1 Total Acidity

3.6.4 Volatile Acidity (Amerine *et al.*, 1980)

Reagents:

a) Sodium hydroxide (0.1 N)
b) Phenolphthalein indicator (1%)

Procedure:

For volatile acidity estimation the apparatus required is a 1000ml wide-mouthed boiling flask, a cylindrical glass tube fitted inside the boiling flask by a wide rubber stopper and Liebig condenser connected vertically to the glass tube. About 500 – 600 ml of distilled water was taken in the flask and heated to boiling for 2-3 minutes. Five milliliters of sample was introduced into the inner tube using a pipette. The rubber stopper connecting the inner tube to the condenser was replaced. The contents were boiled and over 100 ml of distillate was collected in 250 ml conical flask. To the distillate 2-3 drops of phenolphthalein indicator was added. The resultant solution was titrated to a light pink color using 0.1 N NaOH.

Calculations:

1 ml of 0.1 N NaOH = 0.006 g of acetic acid

$$\text{Volatile acidity (\% w/v)} = \frac{\text{Vol of 0.1 N NaOH used x 0.006 x 100}}{\text{Vol of sample taken}}$$

Equation 2 Volatile Acidity

3.6.5 Reducing Sugars (Miller, 1959)

Reagents:

a) Dinitrosalicylic acid (DNS) reagent: 10 g of 3,5-dinitrosalicylic acid, 2.0 g phenol and 0.5 g sodium sulfite were dissolved in 500 ml of 1% sodium hydroxide NAOH solution. The volume was then made to one liter with alkaline solution.

b) Sodium-potassium tartrate (Rochelle Salt) solution: 40.0 g of sodium potassium tartrate were dissolved in distilled water and the volume made to 100 ml.

Procedure:

Test-tubes containing 3 ml of samples and 3 ml of DNS reagent in duplicates, were heated for 15 min in a boiling water bath. Then 1 ml of Rochelle salt solution was added to each tube immediately after boiling and the tubes were subsequently cooled to room temperature. The optical density was measured at 575 nm using Bausch and Lomb Spectronic-20 spectrophotometer. A control was prepared for each experiment. Standard curve was plotted by taking 0.1 to 1.0 mg ml^{-1} concentrations of glucose.

3.6.6 Ethanol Estimation (Caputi et al., 1968)

Reagents:

a) Potassium Dichromate solution: To 200 ml distilled water add 16.8 g potassium dichromate and dissolve it. Then place it in a water bath and add 162.5 ml concentrated sulphuric acid H_2SO_4 slowly. Keep cooling side by side. Make the volume to 500 ml.

Procedure:

For determining the ethanol content of the fermented wash, 1 ml of the sample was taken in 500 ml Pyrex distillation flask and 30 ml of distilled water was added to it. The distillate (20 ml) was collected in 50 ml conical flask containing 25 ml of potassium dichromate $K_2Cr_2O_7$ solution. The flasks were incubated in a water bath maintained at 60 °C for 20 min. They were then cooled to room temperature and the volume made to 50 ml with distilled water. To 5 ml of the distillate collected in potassium dichromate solution, 5 ml of distilled water was added. The optical density was recorded at 600 nm using Bausch and Lomb Spectronic-20 spectrophotometer followed by a control.

A standard curve was prepared by using standard solutions of ethanol containing 2, 4, 6, 8, 10 and 12% (v/v) ethanol in distilled water.

Calculations:

$$\text{Fermentation efficiency (\%)} = \frac{\text{Actual recovery}}{\text{Theoretical recovery}} \text{ x } 100$$

$$\text{Theoretical Recovery} = \text{Total Fermentable Sugars x } 0.51$$

$$\text{Actual Recovery} = \text{Actual ethanol produced}$$

Equation 3 Ethanol Estimation

3.7 Production of Wine

Ginger (*Zingiber officinale*) was used as a substrate for wine production.

3.7.1 Extraction of Juice

The ginger was first peeled and then washed properly with tap water and finally with hot water. All utensils were also washed thoroughly with boiling water before use. The ginger was crushed in a juicer and the must was also added to the juice. The ginger mash was standardized by adding boiled water. The mash was analyzed for total soluble solids, reducing sugars, pH, total acids and volatile acids. Cane sugar syrup was added to the mash to adjust its total soluble solids.

Wheat grains and raisins were washed thoroughly, first with tap water and then with boiling water. The potatoes were peeled and cut into thin slices. These were washed and then treated in boiling water for 5-6 minutes.

3.7.2 Preparation of Starter culture

A loopful of 20 h old actively growing culture of the adapted strain of *S. cerevisiae-220* was used to inoculate 100 ml of glucose yeast extract broth in 250 ml capacity Erlenmeyer flasks. The flasks were incubated at 38 ± 1 °C for 20 h on a rotary shaker (120 rpm). The inoculum thus prepared was used to inoculate a part of the ginger mash which was then incubated at 38 ± 1 °C for 24 h during which vigorous fermentation began. The contents were used as starter culture to carry out the ethanol fermentation.

3.7.3 Fermentation of Ginger mash

The fermentation was carried out in a fermentation vat. The total soluble solids were adjusted to 20.0 °Brix initially with the help of cane sugar syrup using Erma hand refractometer. The pH was adjusted to 5.0 with lemon and orange. Starter culture was added and the fermentation was carried out at 38 ± 1 °C. The contents were shaken 3-4 times a day. The fall in total soluble solids was recorded at regular intervals till steady state values were obtained.

3.7.4 Clarification of Wine

The sediment, including yeast cells, ginger must, raisins, potatoes and wheat, was separated from the wine. The wine was stored at 4 °C to facilitate sedimentation of suspended matter. It was siphoned periodically until there was no sediment.

3.7.5 Wine Analysis and Evaluation

The ginger wine was analyzed for its ethanol content, total soluble solids, reducing sugars, pH, total acids and volatile acids. The wine was also subjected to sensory evaluation using the standard wine score card.

Chapter

4. RESULTS AND DISCUSSION

The results of all the experiments of the present study on "Improvements in efficiency of yeasts for ethanol production" presented in Tables 4.1 to 4.16 have been discussed and summarized accordingly as under:

4.1 Screening of Yeast Strains for Ethanol Production

Ethanol fermentation by eight strains of yeasts, viz. *Saccharomyces cerevisiae-220, S. cerevisiae-360, S. cerevisiae MTCC-181/3c, Trichosporon beigelli, Candida haemulonii, C. parapsilosis, Candida sp.-3(1)* and *Candida sp.-5(2)* has been studied at 30 ± 1 °C, 38 ± 1 °C and at room temperature (37-42 °C). The initial total soluble solids (TSS) in each experiment have been adjusted with cane sugar to 15.0 °Brix. The cane sugar mash (pH 5.0) was inoculated separately with each yeast culture at the rate of 15.0% (v/v) for screening of the yeast strains for ethanol production.

It has been observed from the results presented in Table 4.1 that *S. cerevisiae-220,* fermented more than 50.0% of the substrate within 18 h of fermentation at 30 ± 1 °C while it utilized the same amount in 60 and 72 h at 38 ± 1 °C and at room temperature (37-42 °C) respectively. Fermentation was complete in 60 h at 30 ± 1 °C utilizing substrate to its maximum (84.0%). The substrate utilization by *S. cerevisiae-220* has been maximum (63.0% and 58.0%) when the fermentation was continued for 72 h and 84 h at 38 ± 1 °C and at room temperature (37-42 °C) respectively.

Table 4.1 Ethanol Fermentation by *S. cerevisiae-220* at different Temperatures

Fermentation Period (h)	Fermentation Temperature (°C)		
	30 ± 1	38 ± 1	Room Temperature (37 – 42)
	TSS (°Brix)		
0	15.0	15.0	15.0
12	8.8	12.2	12.6
18	6.5	10.8	12.0
24	5.2	10.5	11.3
36	4.8	9.4	10.7
42	3.3	9.0	9.6
48	2.8	8.3	8.5
60	2.4	7.1	7.8
72	2.4	5.5	7.0
84	2.4	5.5	6.3
96	2.4	5.5	6.3

Substrate Utilization (%)	Fermentation Period (h)		
≥ 50	18	60	72
≥ 75	42	NA	NA
Maximum h (%)	60 (84.0)*	72 (63.3)*	84 (58.0)*

Fermentation Conditions:

Substrate: Cane Sugar
pH: 5.0
Inoculum (v/v): 15.0%
*Maximum substrate utilized %
NA = Maximum substrate utilization < 75.0%

In the experiment conducted with *S. cerevisiae-360* it has been observed that more than 50.0% of substrate was utilized in 42 h at 30 ± 1 °C (Table 4.2). the period for fermenting half the amount of the total substrate was recorded to be more at 38 ± 1 °C as well as at room temperature, (48 h and 60 h) respectively. Substrate was utilized to its maximum (80.0%) during 84 h at 30 ± 1 °C, while maximum substrate utilization has been less than 70.0% at 38 ± 1 °C as well as at room temperature. Substrate utilization decreased considerably with the increase in fermentation temperature from 30 ± 1 °C to 38 ± 1 °C.

Table 4.2 Ethanol Fermentation by *S. cerevisiae-360* at different Temperatures

Fermentation Period (h)	Fermentation Temperature (°C)		
	30 ± 1	38 ± 1	Room Temperature (37 – 42)
	TSS (°Brix)		
0	15.0	15.0	15.0
12	12.6	12.8	12.8
18	11.2	12.1	12.0
24	9.0	10.4	11.0
36	8.2	9.0	10.2
42	7.4	7.9	9.3
48	6.4	6.8	7.8
60	5.1	6.5	7.0
72	4.0	4.6	6.0
84	3.0	4.6	5.4
96	3.0	4.6	5.4

Substrate Utilization (%)	Fermentation Period (h)		
≥ 50	42	48	60
≥ 75	84	NA	NA
Maximum h (%)	84 (80.0)*	72 (69.3)*	84 (64.0)*

Fermentation Conditions:

Substrate: Cane Sugar
pH: 5.0
Inoculum (v/v): 15.0%
*Maximum substrate utilized %
NA = Maximum substrate utilization < 75.0%

S. cerevisiae MTCC-181/3c fermented 88.0% substrate in 96 h at 30 ± 1 °C. from the observations of all the strains screened, it has been seen that this strain utilized maximum amount of substrate (88.0%). But it has been observed to be slower in substrate utilization than *S. cerevisiae-220* which fermented 56.6% of cane sugar in 18 h at 30 ± 1 °C (Table 4.1), whereas *S. cerevisiae MTCC-181/3c* utilized 53.3% of substrate during 36 h at the same temperature (Table 4.3). Maximum substrate utilization by *S. cerevisiae MTCC-181/3c* decreased to 63.3% and 59.3% respectively, at 38 ± 1 °C and at room temperature.

Table 4.3 Ethanol Fermentation by *S. cerevisiae MTCC-181/3c* at different Temperatures

Fermentation Period (h)	Fermentation Temperature (°C)		
	30 ± 1	38 ± 1	Room Temperature (37 – 42)
	TSS (°Brix)		
0	15.0	15.0	15.0
12	9.8	12.5	12.8
18	9.0	11.2	11.6
24	8.3	10.0	11.1
36	7.0	9.0	10.5
42	6.0	8.4	9.2
48	5.6	7.9	8.3
60	4.8	6.6	7.1
72	4.5	6.2	6.5
84	3.0	5.8	6.1
96	1.8	5.5	6.1

Substrate Utilization (%)	Fermentation Period (h)		
≥ 50	36	60	60
≥ 75	84	NA	NA
Maximum h (%)	96 (88.0)*	96 (63.3)*	84 (59.3)*

Fermentation Conditions:

Substrate: Cane Sugar
pH: 5.0
Inoculum (v/v): 15.0%
*Maximum substrate utilized %
NA = Maximum substrate utilization < 75.0%

The results of ethanol fermentation by *Trichosporon beigelli* at 30 ± 1 °C, 38 ± 1 °C and at room temperature (37-42 °C) have been presented in Table 4.4. This yeast culture has not been found to be efficient in cane sugar fermentation since it utilized 40.0% or less than 40.0% of the total substrate in 96 h at all the three fermentation temperatures maintained to study the effect of temperature on fermentation.

Table 4.4 Ethanol Fermentation by *Trichosporon beigelli* at different Temperatures

Fermentation Period (h)	Fermentation Temperature (°C)		
	30 ± 1	38 ± 1	Room Temperature (37 – 42)
	TSS (°Brix)		
0	15.0	15.0	15.0
12	11.6	14.1	14.2
18	11.1	13.8	13.7
24	10.8	13.0	13.0
36	10.6	12.4	12.6
42	10.4	11.5	12.0
48	10.0	10.7	11.5
60	9.8	10.3	10.9
72	9.5	10.0	10.6
84	9.3	9.6	10.4
96	9.0	9.4	10.0

Substrate Utilization (%)	Fermentation Period (h)		
≥ 25	18	48	60
≥ 50	NA	NA	NA
Maximum h (%)	96 (40.0)*	96 (37.3)*	96 (33.3)*

Fermentation Conditions:

Substrate: Cane Sugar
pH: 5.0
Inoculum (v/v): 15.0%
*Maximum substrate utilized %
NA = Maximum substrate utilization < 50.0%

The results of the experiments of ethanol fermentation by the four *Candida* species have been presented in Tables 4.5 to 4.8. In each experiment it has been observed that less than 50.0% of the total substrate has been fermented when the temperatures 30 ± 1 °C, 38 ± 1 °C and room temperature (37-42 °C) were maintained for fermentation. It has been only *C. haemulonii* which fermented 50.0% of cane sugar in 96 h at 30 ± 1 °C (Table 4.5). Fermentation got completely stopped after 48 h at room temperature and during this period it had utilized just 10.0% of total substrate.

Table 4.5 Ethanol Fermentation by *Candida haemulonii* at different Temperatures

Fermentation Period (h)	Fermentation Temperature (°C)		
	30 ± 1	38 ± 1	Room Temperature (37 – 42)
	TSS (°Brix)		
0	15.0	15.0	15.0
12	13.0	13.2	14.6
18	12.6	12.8	14.4
24	12.2	12.3	14.2
36	11.8	11.9	14.0
42	11.6	10.8	13.8
48	11.2	10.4	13.5
60	9.0	9.8	13.5
72	8.3	9.0	13.5
84	7.8	8.6	13.5
96	7.5	8.2	13.5

Substrate Utilization (%)	Fermentation Period (h)		
≥ 25	48	42	NA
≥ 50	96	NA	NA
Maximum h (%)	96 (50.0)*	96 (45.3)*	48 (10.0)*

Fermentation Conditions:

Substrate: Cane Sugar
pH: 5.0
Inoculum (v/v): 15.0%
*Maximum substrate utilized %
NA = Maximum substrate utilization < 50.0%

Candida parapsilosis utilized maximum substrate (46.6%, 39.3% and 36.0%) in 96 h at 30 ± 1 °C, 38 ± 1 °C and room temperature (37-42 °C), respectively (Table 4.6).

Table 4.6 Ethanol Fermentation by *Candida parapsilosis* at different Temperatures

Fermentation Period (h)	Fermentation Temperature (°C)		
	30 ± 1	38 ± 1	Room Temperature (37 – 42)
	TSS (°Brix)		
0	15.0	15.0	15.0
12	13.0	13.2	14.6
18	12.6	12.8	14.4
24	12.2	12.3	14.2
36	11.8	11.9	14.0
42	11.6	10.8	13.8
48	11.2	10.4	13.5
60	9.0	9.8	13.5
72	8.3	9.0	13.5
84	7.8	8.6	13.5
96	7.5	8.2	13.5

Substrate Utilization (%)	Fermentation Period (h)		
≥ 25	48	42	NA
≥ 50	96	NA	NA
Maximum h (%)	96 (50.0)*	96 (45.3)*	48 (10.0)*

Fermentation Conditions:

Substrate: Cane Sugar
pH: 5.0
Inoculum (v/v): 15.0%
*Maximum substrate utilized %
NA = Maximum substrate utilization < 50.0%

Candida sp.-3(1) also showed a decrease in the percentage of total substrate fermented as the temperature was raised, the maximum substrate utilization being 46.6%, 37.3% and 26.6% in 96 h at 30 ± 1 °C, 38 ± 1 °C and room temperature (37-42 °C), respectively (Table 4.7).

Table 4.7 Ethanol Fermentation by *Candida sp.-3(1)* at different Temperatures

Fermentation Period (h)	Fermentation Temperature (°C)		
	30 ± 1	38 ± 1	Room Temperature (37 – 42)
	TSS (°Brix)		
0	15.0	15.0	15.0
12	12.8	13.0	14.6
18	12.3	12.6	14.2
24	11.6	12.1	13.8
36	11.0	11.8	13.6
42	9.8	11.4	13.3
48	9.5	10.8	12.8
60	9.0	10.5	11.7
72	8.7	10.0	11.5
84	8.2	9.7	11.2
96	8.0	9.4	11.0

Substrate Utilization (%)	Fermentation Period (h)		
≥ 25	36	48	84
≥ 50	NA	NA	NA
Maximum h (%)	96 (46.6)*	96 (37.3)*	96 (26.6)*

Fermentation Conditions:

Substrate: Cane Sugar
pH: 5.0
Inoculum (v/v): 15.0%
*Maximum substrate utilized %
NA = Maximum substrate utilization < 50.0%

Candida sp.-5(2) has been found to ferment greater amount of substrate at 38 ± 1 °C (46.6%) than at 30 ± 1 °C (41.3%). But maximum substrate utilization decreased at room temperature (37-42 °C) to 37.3%. The fermentation completed in 96 h in three sets maintained at different temperatures is given in Table 4.8.

Keeping in view the results presented in Tables 4.1 to 4.8, there has been observed a decrease in substrate utilization by all the yeast strains, except *Candida sp.-5(2)*, at 38 ± 1 °C as well as at room temperature (37-42 °C) as compared to that at 30 ± 1 °C. So, for all these strains 30 ± 1 °C is the optimum temperature for fermentation. *S. cerevisiae-360* utilized more of the substrate (69.3%) as compared to *S. cerevisiae-220* (63.3%) at 38 ± 1 °C. Taking into account this property of *S. cerevisiae-360* to ferment the substrate efficiently at elevated fermentation temperature, it was screened for further studies for protoplast fusion to obtain a strain with higher ethanol fermentation at higher temperature.

Richter and Becker (1987) have reported that maximum specific ethanol formation rate can be reached within the temperature limits of 32 – 36 °C. Calazans *et al.*, (1990) reported that ethanol yield was not affected by temperature in the range of 30.0 – 37.5 °C for alcoholic fermentation by *Zymomonas mobilis ZAP*, though the optimum temperature has been 35 °C. they have observed a decrease in lag phase at 35 – 40 °C. Neelam and Amarjit (1991) screened various yeast cultures for ethanol production at 37 °C and reported that the sugar conversion efficiency of the yeast strains varied from 66.0% to 88.5%. according to Groot *et al.*, (1992), a temperature increase from 30 °C in the fermenter to 35 °C or more in the recycle loop led to a significantly lower ethanol concentration in the broth.

It may be concluded from the results presented in Tables 4.1 to 4.8 that out of the eight strains of yeast screened for ethanol fermentation at different temperatures, only *S. cerevisiae-360* could utilize the substrate efficiently at elevated temperature, i.e., 38 ± 1 °C.

Table 4.8 Ethanol Fermentation by *Candida sp.-5(2)* at different Temperatures

Fermentation Period (h)	Fermentation Temperature (°C)		
	30 ± 1	38 ± 1	Room Temperature (37 – 42)
	TSS (°Brix)		
0	15.0	15.0	15.0
12	13.8	13.0	14.6
18	13.4	12.5	14.0
24	12.9	12.0	13.8
36	12.0	11.7	13.4
42	11.0	11.0	12.5
48	10.6	9.8	12.0
60	10.0	9.4	11.6
72	9.8	8.8	10.4
84	9.4	8.5	10.0
96	8.8	8.0	9.4

Substrate Utilization (%)	Fermentation Period (h)		
≥ 25	42	42	72
≥ 50	NA	NA	NA
Maximum h (%)	96 (41.3)*	96 (46.6)*	96 (37.3)*

Fermentation Conditions:

Substrate: Cane Sugar
pH: 5.0
Inoculum (v/v): 15.0%
*Maximum substrate utilized %
NA = Maximum substrate utilization < 50.0%

4.2 Temperature Tolerance Profile Studies on *S. cerevisiae*

Saccharomyces cerevisiae-220 and *Saccharomyces cerevisiae-360* have been compared for their growth in glucose yeast extract broth by recording the optical density of the broth at 540 nm at 30 ± 1 °C and 38 ± 1 °C. The results presented in Tables 4.9 and 4.10 indicate that the lag phase has been only for a period of 4 h before the initiation of exponential phase. Turbidity decreased markedly in both the cultures at 38 ± 1 °C as compared to that at 30 ± 1 °C. The initiation of stationary phase was recorded after 24 h of incubation when temperatures of 30 ± 1 °C and 38 ± 1 °C were maintained for the growth of *S. cerevisiae-220*, whereas, the log phase prolonged up to 30 h at 38 ± 1 °C, after which the stationary phase started in experiments with *S. cerevisiae-360*. There has been an increase in growth of *S. cerevisiae-220* as compared to that of *S. cerevisiae-360* at 30 ± 1 °C as well as at 38 ± 1 °C. This increase in growth of *S. cerevisiae-220* may be attributed to faster multiplication of this strain as compared to *S. cerevisiae-360* at both 30 ± 1 °C and 38 ± 1 °C.

Barillere *at al.*, (1985) obtained different types of survival curves in a study of heat resistance of different *Saccharomyces* species. Yamamura *et al.*, (1988) reported an increase in growth of *Saccharomyces cerevisiae* at 40 °C even with a smaller inoculum size (10^5 cells ml^{-1}) when supplemented with yeast extract (6%).

It may be summarized from the results of the experiments conducted to study the effect of temperature on growth of *S. cerevisiae-220* and *S. cerevisiae-360* that *S. cerevisiae-220* could grow more efficiently than *S. cerevisiae-360* at 30 ± 1 °C as well as at 38 ± 1 °C.

Table 4.9 Effect of Temperature on growth of *S. cerevisiae-220*

Incubation Period (h)	Incubation Temperature (°C)	
	30 ± 1	38 ± 1
	Optical Density[1] (at 540 nm)	
0	0.06	0.08
2	0.06	0.08
4	0.06	0.08
6	0.15	0.16
8	0.28	0.31
10	0.42	0.46
12	0.52	0.58
14	0.78	0.80
16	0.96	1.00
18	1.18	1.20
20	1.28	1.29
22	1.39	1.33
24	1.45	1.35
26	1.46	1.36
28	1.46	1.36
30	1.46	1.36
32	1.46	1.36

Conditions for Growth:

Medium: GYE Broth

pH: 5.0

[1] Optical Density is a method to quantify parameters like concentration of cells, production of biomass, etc. It is a dimensionless quantity and does not have any scientific unit. The expression for optical density is given by Log10 (1/T), where T is the transmittance. Since optical density is inversely proportional to transmittance, higher the optical density, lower will be the transmittance of light waves through the biomass.

Table 4.10 Effect of Temperature on growth of *S. cerevisiae-360*

Incubation Period (h)	Incubation Temperature (°C)	
	30 ± 1	38 ± 1
	Optical Density (at 540 nm)	
0	0.07	0.09
2	0.07	0.09
4	0.07	0.09
6	0.09	0.11
8	0.20	0.21
10	0.43	0.37
12	0.61	0.58
14	0.81	0.79
16	1.00	0.91
18	1.15	1.04
20	1.22	1.06
22	1.29	1.13
24	1.32	1.16
26	1.34	1.18
28	1.34	1.20
30	1.34	1.22
32	1.34	1.22

Conditions for Growth:

Medium: GYE Broth
pH: 5.0

4.3 Ethanol Tolerance Profile Studies on *S. cerevisiae*

The ability of the fermenting organism to tolerate ethanol is one of the key constraints in achieving high ethanol fermentation efficiency to keep the fermentation capital cost low (Righelato, 1980). It has long been recognized that industrial ethanol fermentation terminates prematurely despite an unlimited supply of substrate. This premature termination of fermentation limits the final accumulation of ethanol that can be achieved.

The fermenting ability is related to viability, growth and fermentation in the presence of varying concentrations of ethanol. So, the ethanol tolerance of *S. cerevisiae-220* and *S. cerevisiae-360* was studied in terms of growth (optical density of 540 nm) in glucose yeast extract broth, initially supplemented with ethanol concentration ranging from 10.0 to 15.0% (v/v) for *S. cerevisiae-220* and 8.0 to 15.0% (v/v) for *S. cerevisiae-360* at 30 ± 1 °C.

Supplementation of ethanol at varying concentrations in the growth medium decreased the optical density in both the strains. (Tables 4.11 and 4.12). The optical density of samples has been observed after 12 h of inoculation of substrate.

An increase in optical density up to 24 h with 10.0% and 11.0% (v/v) initial concentration of ethanol has been observed with *S. cerevisiae-220.* The optical density has been much less than that of control treatment. Optical density has been constant after 24 h of incubation. Growth was highly suppressed at 12.0% and 13.0% (v/v) initial alcohol concentrations. No change has been recorded in the optical density after 12 h at 14.0% and 15.0% (v/v) alcohol concentrations. It has been observed that cells of *S. cerevisiae-220* remain viable/active when the growth medium was initially supplemented with 13.0% (v/v) of ethanol.

The results of the effect of ethanol on the growth of *S. cerevisiae-360* presented in Table 4.12 indicates that the optical density increased constantly when ethanol (8.0% v/v) was supplemented in the growth medium though it was much less as compared to the control treatment. A gradual decrease in optical density has been observed from 12 h onwards up to 40 – 42 h with initial ethanol concentration 10.0, 12.0, 13.0, 14.0 and 15.0% (v/v). Thereafter a slight increase in turbidity has been observed, which may be attributed to the adaptation of the yeast cells to the respective ethanol concentrations. *S. cerevisiae-360* tolerated supplementation of ethanol @8.0% (v/v) in the medium, as the growth of this strain decreased gradually in the medium supplemented with higher concentrations of ethanol.

Taking into consideration the property of *S. cerevisiae-220* to grow in 13.0% (v/v) ethanol concentration supplemented in the growth medium, this strain was screened as the ethanol-tolerant strain and used for studies on protoplast fusion to construct a strain capable of fermentation at high concentrations of ethanol at elevated temperature.

The variation in ethanol tolerance among the strains screened for ethanol tolerance could be attributed to the fact that all strains of *Saccharomyces* are not equally tolerant, and ethanol resistance is determined in part by genetic composition (Gray, 1941). Ismail and Ali (1971) also reported that yeast strains differ in their alcohol tolerance. *Saccharomyces* species have been reported to grow in a medium supplemented with 8 – 12% (v/v) ethanol and survive an exposure to about 15% (v/v) (Rose, 1980, 1983). Brown *et al.,* (1981) observed an immediate reduction in growth rate with addition of ethanol at 4% and 8% (v/v) to the log phase cultures of *S. cerevisiae*, however they reported no effects on growth at ethanol concentrations below 1.2% (v/v). The growth was not completely inhibited up to a concentration of 12.5% (v/v). Whereas, Sipiczki *et al.,* (1988) have reported that some strains of *Saccharomyces* could grow intensively in 12.0% (v/v) ethanol but did not show high productivity. Goyal (1989) observed growth of *S. cerevisiae* up to 11.0% initial ethanol concentration. According to the results of El-Diwany *et al.,* (1992), the yeast tolerated 10% ethanol but failed to grow at ethanol concentration varying from 12 to 15%.

Table 4.11 Effect of Ethanol on growth of *S. cerevisiae-220*

Incubation Period (h)	Alcohol Concentration (% v/v)						
	Control	10.0	11.0	12.0	13.0	14.0	15.0
	Optical Density (at 540 nm)						
12	0.52	0.58	0.56	0.41	0.30	0.25	0.22
14	0.78	0.66	0.60	0.43	0.35	0.25	0.22
16	0.96	0.68	0.62	0.44	0.38	0.25	0.22
18	1.18	0.77	0.66	0.46	0.40	0.25	0.22
20	1.28	0.83	0.69	0.48	0.42	0.25	0.22
22	1.39	0.85	0.72	0.50	0.42	0.25	0.22
24	1.45	0.86	0.78	0.54	0.42	0.25	0.22
36	1.46	0.86	0.80	0.57	0.42	0.25	0.22
38	1.46	0.86	0.80	0.57	0.42	0.25	0.22
40	1.46	0.86	0.80	0.57	0.42	0.25	0.22
42	1.46	0.86	0.80	0.57	0.42	0.25	0.22
44	1.46	0.86	0.80	0.57	0.42	0.25	0.22
46	1.46	0.86	0.80	0.57	0.42	0.25	0.22

Growth Conditions:

Growth Medium: GYE Broth
Temperature: 30 ± 1 °C
pH: 5.0

Table 4.12 Effect of Ethanol on growth of *S. cerevisiae-360*

Incubation Period (h)	Alcohol Concentration (% v/v)						
	Control	8.0	10.0	12.0	13.0	14.0	15.0
	Optical Density (at 540 nm)						
12	0.61	0.29	0.15	0.13	0.12	0.12	0.10
14	0.81	0.35	0.14	0.12	0.11	0.11	0.09
16	1.00	0.42	0.13	0.12	0.11	0.11	0.08
18	1.15	0.53	0.10	0.11	0.10	0.11	0.08
20	1.22	0.64	0.12	0.10	0.10	0.10	0.08
22	1.29	0.72	0.12	0.10	0.09	0.10	0.07
24	1.32	0.82	0.11	0.09	0.08	0.09	0.07
36	1.34	0.85	0.10	0.08	0.08	0.09	0.06
38	1.34	0.87	0.09	0.07	0.07	0.08	0.05
40	1.34	0.88	0.10	0.07	0.06	0.07	0.04
42	1.34	0.90	0.15	0.10	0.05	0.07	0.04
44	1.34	0.91	0.15	0.11	0.10	0.11	0.07
46	1.34	0.92	0.15	0.12	0.11	0.12	0.07

Growth Conditions:

Growth Medium: GYE Broth
Temperature: 30 ± 1 °C
pH: 5.0

4.4 Protoplast Studies on Yeast

The experiments on protoplast fusion of yeast have been conducted to develop a yeast strain which can withstand high concentration of ethanol and also ferment the substrate at elevated temperature. Protoplast fusion has been carried out between *Saccharomyces cerevisiae-220* and *Saccharomyces cerevisiae-360* which have been screened as the ethanol-tolerant strain (13.0% v/v) and utilizing substrate with greater efficiency at higher temperature (38 ± 1 °C).

4.4.1 Protoplast Isolation

The protoplasts from the cells of *S. cerevisiae-220* and *S. cerevisiae-360* have been isolated by the method of Dickinson and Isenberg (1982). Cells from exponential phase have been subjected to lysis by Novozym 234.

4.4.2 Viability of Protoplasts

The viability of the protoplasts of *S. cerevisiae-220* and *S. cerevisiae-360* have been thoroughly tested separately by spreading them on glucose yeast extract medium supplemented with sorbitol (0.6 M), as osmotic stabilizer, followed by a control i.e., without sorbitol supplementation. The plates were incubated at 30 ± 1 °C and 38 ± 1 °C for *S. cerevisiae-220* and *S. cerevisiae-360* respectively. It has been observed that the protoplasts could not regenerate in the medium without osmotic stabilizer, but they formed colonies in the medium containing sorbitol.

4.4.3 Protoplast Fusion and Regeneration

Protoplast suspensions, one ml each of *S. cerevisiae-220* and *S. cerevisiae-360* were mixed together and centrifuged at 3000 rpm for 2 min. Polyethylene glycol (34%) containing sorbitol (0.8 M) and calcium chloride (10 mM) was added to the pellet obtained. The suspension was incubated at 30 ± 1 °C for 50 min. Fusogen polyethylene glycol helped in fusion of the protoplasts. The protoplasts were subsequently washed with buffer (SKM) and centrifuged at 3000 rpm for 2 min. The aggregated protoplasts diluted with buffer (SKM) were spread on plates containing glucose yeast extract agar medium supplemented with sorbitol (0.8 M). The plates were incubated at 38 ± 1 °C for 48 h. It was observed that a number of colonies were formed during this period of incubation.

Some of the colonies mentioned as above were selected at random and subcultured on glucose yeast extract medium. The plates containing the selected colonies have been incubated at 38 ± 1 °C for 48 h. The same were maintained on glucose yeast extract slants at 4 °C for further studies.

The above referred colonies have been screened for their efficiency to ferment cane sugar at 38 ± 1 °C. It has been observed that none of them could ferment the substrate with higher efficiency as compared to their parent strains, i.e., *S. cerevisiae-220* and *S. cerevisiae-360.*

Farahnak *et al.,* (1986) constructed a yeast strain by fusing protoplasts of lactose fermenting *Kluyveromyces fragilis* and high ethanol-tolerant *S. cerevisiae* and thus obtained a hybrid which could ferment the substrate at higher concentration of ethanol. Laluce *et al.,* (1989) improved the thermotolerance of amylolytic yeasts by protoplast fusion between petite mutants of *Saccharomyces diastaticus* and a strain of *Saccharomyces cerevisiae* which can grow and ferment at 40 °C. Kida *et al.,* (1992) constructed a thermotolerant flocculating yeast *S. cerevisiae-KF-7* by protoplast fusion of the flocculating yeast *S. cerevisiae-IR-2* and the thermotolerant yeast *S. cerevisiae-EP-1.* The hybrid fermented at 35 °C with an ethanol productivity of 3.6 g/l/h.

It may be concluded from the results of the experiments on protoplast fusion of *S. cerevisiae*-220 and *S. cerevisiae-360* that yeast strains obtained and screened as mentioned above could not tolerate higher concentration of ethanol and also could not ferment the substrate at elevated temperature (38 ± 1 °C).

4.5 Adaption of *S. cerevisiae-220* to Higher Temperature

The ethanol-tolerant yeast, *S. cerevisiae-220* was subjected to its adaptation to higher temperature up to 38 ± 1 °C. The yeast strain was subcultured for ten successive transfers on glucose yeast extract slants and incubated at 38 ± 1 °C for 48 h. The purity of the culture was again ascertained after the tenth successive transfer by plating the culture on growth medium. The successively adapted strain of yeast *S. cerevisiae-220* has been labelled as *S. cerevisiae-220A*. This strain has been screened for the production of ethanol by utilizing cane sugar and subsequently for the production of ginger wine at 38 ± 1 °C.

Table 4.13 Ethanol Fermentation by *S. cerevisiae-220A at* 38 ± 1 °C

Fermentation Period (h)	**TSS (°Brix)**
0	15.0
12	10.6
18	9.4
24	8.7
36	8.0
42	7.4
48	6.2
60	5.3
72	5.3
84	5.3
96	5.3

Substrate Utilization (%)	**Fermentation Period (h)**
≥ 50	42
≥ 75	NA
Maximum	60 (64.6%)*

Fermentation Conditions:

Substrate: Cane Sugar
pH: 5.0
Inoculum (v/v): 15.0%
*Maximum substrate utilized %
NA = Maximum substrate utilization < 75.0%

The results of the ethanol fermentation by *S. cerevisiae-220A* have been presented in Table 4.13. Substrate (50.0%) has been utilized in 42 h by *S. cerevisiae-220A* whereas *S. cerevisiae-220* utilized 50.0% of the substrate in 60 h (see Table 4.1). *S. cerevisiae-220* utilized substrate to its maximum (63.3%) in 72 h while *S. cerevisiae-220A* completed the fermentation in 60 h with maximum substrate utilization (64.6%).

It can be summarized from the results of the experiment conducted on ethanol production by *S. cerevisiae-220A* that the training method has successfully enabled *S. cerevisiae-220* to utilize substrate more efficiently at 38 ± 1 °C. It can also be concluded that the lag phase of fermentation has been significantly curtailed in the case of *S. cerevisiae-220A* as compared to *S. cerevisiae-220* at 38 ± 1 °C.

4.6 Production of Ginger Wine

The ginger mash has been analyzed for its total soluble solids, reducing sugars, total acids, volatile acids and pH and the results have been presented in Table 4.14. The ingredients used and the conditions maintained for the production of wine from *Zingiber officianale* have been presented in Table 4.15. *Saccharomyces cerevisiae-220A* has been used for the production of ginger wine.

The fermentation was completed within 72 h and the wine was siphoned periodically thereafter. The ginger wine was stored at 4 °C. the suspended matter was removed by decantation. The clarified wine was bottled and analyzed. The results of the analysis of ginger wine, i.e., ethanol concentration, total soluble solids, reducing sugars, total acids, volatile acids and pH have been presented in Table 4.16. The wine has been subjected to sensory evaluation using the wine score card and has been graded as commercially outstanding wine.

Table 4.14 Analysis of Ginger Mash

Total Soluble Solids (TSS, °Brix)	3.0
Reducing Sugars (g%)	0.18
Total Acids (% Tartaric Acid)	0.26
Volatile Acids (% Acetic Acid)	0.22
pH	5.2

Table 4.15 Production of Ginger Wine

Ingredients	
Ginger	4 kg
Raisins	1 kg
Potato	1 kg
Wheat	0.5 kg
Lemon	4 pieces
Orange	2 pieces
Water	10 liters
Conditions	
Initial TSS (°Brix)	20.0
pH	5.0
Temperature (°C)	38 ± 1

Table 4.16 Analysis of Ginger Wine

Ethyl Alcohol (% v/v)	10.5
Total Soluble Solids (TSS, °Brix)	3.0
Reducing Sugars (g%)	0.11
Total Acids (% Tartaric Acid)	0.34
Volatile Acids (% Acetic Acid)	0.24
pH	4.3
Grading	Commercially Outstanding

Chapter

5. SUMMARY

Eight strains of yeasts, namely

1) *S. cerevisiae-220,*
2) *S. cerevisiae-360,*
3) *S. cerevisiae-181/3c,*
4) *Trichosporon beigelli,*
5) *Candida haemulonii,*
6) *C. parapsilosis,*
7) *Candida sp.-3(1)* and
8) *Candida sp.-5(2)*

were evaluated for their potential to ferment cane sugars to ethanol at 30 ± 1 °C, 38 ± 1 °C and at room temperature (37-42 °C). Utilization of the substrate at all these temperatures varied widely. Fermentation period ranged form 60 h to 90 h. Substrate utilization by all strains decreased at 38 ± 1 °C and at room temperature as compared to 30 ± 1 °C.

T. beigelli, C. haemulonii, C. parapsilosis, Candida sp.-3(1) and *Candida sp.-5(2)* utilized not more than 50.0 percent of the substrate.

C. haemulonii utilized 50.0% (maximum) at 30 ± 1 °C whereas 10% (minimum) at room temperature.

S. cerevisiae-220 utilized maximum amount of substrate (84.0%) in minimum time (60 h) at 30 ± 1 °C as compared to 88.0% by *S. cerevisiae-181/3c* in 96 h at the same temperature.

S. cerevisiae-360 utilized substrate more efficiently (69.3%) at 38 ± 1 °C followed by *S. cerevisiae-220* (63.3%). *S. cerevisiae-220* multiply faster at 30 ± 1 °C as well as at 38 ± 1 °C as compared to *S. cerevisiae-360. S. cerevisiae-220* could withstand 13.0% ethanol whereas *S. cerevisiae-360* could tolerate only 8.0% of ethanol supplemented in the growth medium.

The protoplast of ethanol-tolerant strain, *S. cerevisiae-220* was fused with that of *S. cerevisiae-360.* None of the colonies formed could ferment cane sugar to ethanol more efficiently than their parents at 38 ± 1 °C. The

adapted strain of *S. cerevisiae-220* at 38 ± 1 °C utilized 64.6% of substrate during 60 h as compared to *S. cerevisiae-220* utilizing 63.3% of substrate in 72 h.

Wine with commercially outstanding quality has been prepared from *Zingiber officianale* with *S. cerevisiae-220-A* at 38 ± 1 °C.

BIBLIOGRAPHY

REFERENCES

*Abate, C.M., Callieri, D.A.S., Acosta, S. and Vie, M.de. 1987. Production of ethanol by a flocculent Saccharomyces sp. in a continuous up-flow reactor using sucrose, sugar-cane juice and molasses as the carbon source. *Mircen J. Appl. Microbiol.Biotechnol.* **3**(4): 401-409.

Abdel-Fattah, A.F., Abouzeid, A.Z.A. and Farid, M.A. 1984. Production of ethyl alcohol by Saccharomyces cerevisiae, including utilization of onion juice. *Agric. Wastes.* **9**(2): 101-110.

*Abdel- Salam, M.S., Abd - El- Halim, M.M., Ghanem, K.M. and El-Kady, M.H.A. 1992. Genetic construction of a Saccharomyces cerevisiae strain capable of lactose assimilation. Bulletin of Faculty of Agriculture, University of Cairo. **43**(1): 87-101.

Abouzeid, M.M. and Reddy, C.A. 1986. Direct fermentation of potato starch to ethanol by cocultures of Aspergillus niger and Saccharomyces cerevisiae. *Appl. Environ. Microbiol.* **52**(5): 1055-1059.

Agrawal, P. and Verramallu, U. 1990. Ethanol fermentation by Zymomonas mobilis ATCC 10988 in repeated batch cultures. *J. Chem. Technol. Biotechnol.* **47**(1):1-14.

Agudo, L. del C. 1985. Genetic aspects of ethanol tolerance and production by Saccharomyces cerevisiae. *Curr. Microbiol.* **12**(1): 41-44.

Agudo, L. del C. 1992. Lipid content of Saccharomyces cerevisiae strains with different degrees of ethanol tolerance. *Appl. Microbiol. Biotechnol.* **37**(5): 647-651.

Akpan, I., Ikenebomeh, M.J., Uraih, N. and Obuekwe, C.O. 1988. Production of ethanol from cassava whey. *Acta. Biotechnol.* **8**(1):39-45.

*Alian, A. and Musenge, H.M. 1970. Utilization of pineapple waste for wine making. *Zambia J. Sci. Technol.* **1**(1): 29-33.

*Ameh, J.B., Okagbue, R.N. and Ahmad, A.A. 1990. Isolation and characterization of local yeast strains for ethanol production. *Nigerian J. Technol. Res.* **1**(1): 47-52.

Amerine, M.A., Berg, H.W. and Cruess, W.V. 1967. Technology of wine making. The AVI Publishing Co. Inc. Evolution of wines and brandies. 678-724.

Amerine, M.A. and Kunkee, R.E. 1968. Microbiology of wine making. *Ann. Rev. Microbiol.* **22**: 323-358.

Amerine, M.A., Kunkee, R.E., Ough, C.S., Singleton, V.L. and Webb, A.D. 1980. The technology of wine making, 4th ed. The AVI Publishing Co. Westport, C.T.

Amerine, M.A. and Singleton, V.L. 1968. Wine: An Introduction For Americans. University of California Press, Berkeley and Los Angeles.

Argiriou, T., Kalliafas, A., Psarianos, C., Kana, K., Kanellaki, M. and Koutinas, A.A. 1992. New alcohol resistant strains of Saccharomyces cerevisiae species for potable alcohol production using molasses. *Appl. Biochem. Biotechnol.* **36**(3): 153-161.

Arthur, H. and Watson, K. 1976. Thermal adaptation in yeast: growth temperatures, membrane lipid and cytochrome composition of psychrophilic, mesophilic and thermophilic yeasts. *J. Bacteriol.* **128**(1): 56-68.

Ballesteros, I., Ballesteros, M., Cabanas, A., Carrasco, J., Martin, C., Negro, M.J., Saez, F. and Saez, R. 1991. Selection of thermotolerant yeasts for simultaneous saccharification and fermentation (SSF) of cellulose to ethanol. *Appl. Biochem. Biotechnol.* **28-29**: 307-315.

Baraniak, A. 1988. Fermentation of D-xylose into ethanol by yeast strains. *Acta Microbiol. Pol.* **37**(2): 227-230.

Bardiya, M.C., Sharma, S. and Tauro, P. 1983. Yeast breeding for higher alcohol production. VIIIth Int. Specialized Symp on Yeast. Bombay. Jan 24-28, 1983. Abstr. of Papers. *Indian J. Microbiol.* **23**: 27.

Barillere, J.M., Mimouni, A., Dubois, C., and Bidan, P. 1985. Heat resistance of strains of yeasts isolated from wine. 1. Effects of the experimental conditions on the shape of survival curves. *Sci. Aliments.* **5**(3): 365-378.

Beach, D. and Nurse, P. 1981. High-frequency transformation of the fission yeast Schizosaccharomyces pombe. *Nature.* **290**: 140-142.

Beaven, M.J., Charpentier, C. and Rose, A.H. 1982. Production and tolerance of ethanol in relation to phospholipid fatty-acyl composition in Saccharomyces cerevisiae NCYC 431. *J. Gen. Microbiol.* **128**: 1447-1455.

Bechard, P., Jolicoeur, C. and Beaubien, A. 1987. Toxicity effects of alcohols on Saccharomyces cerevisiae. A flow micro calorimetry investigation. *Biotechnol. Technol.* **1**(2): 73-78.

Benitez, T., Lucas, D.C., Andres, A., Jaime, C. and Olmedo, C.E. 1983. Selection of wine yeasts for growth and fermentation in presence of ethanol and sucrose. *Appl. Environ. Microbiol.* **45**(5): 1429-36.

Bertolini, M.C., Ernandes, J.R. and Laluce, C. 1991. New yeast strains for alcoholic fermentation at higher sugar concentration. *Biotechnol. Lett.* **13**(3): 197-202.

Bhatt, S., Rana, R.S., and Nain, L.R. 1987. Ethanol production from mixed fruit juice of damaged guava and banana. *J. Food Sci. Technol. India.* **24**(4): 192-193.

Brown, J.P. 1971. Susceptibility of the cell walls of some yeasts to lysis by enzymes of Helix pomatia. *Can. J. Microbiol.* **17**: 205-208.

Brown, S.W., Oliver, S.G. and Harrison, D.E.F. 1981. Ethanol inhibition of yeast growth and fermentation differences in magnitude and complexity of the effect. *Eur. J. Appl. Microbiol. Biotechnol.* **11**(3): 151-155.

*Calazans, G.M.T., Rios, E.M., Morias, J.O.F. de and De- Morais, J.O.F. 1990. Effect of glucose concentration, initial pH, and temperature on the alcoholic fermentation by Zymomonas mobilis ZAP, *Arquivos de-Biologia-e-Technologia.* **33**(2): 439-455.

Caputi, A. Jr., Ueda, M. and Brown, T. 1968. Spectrophotometric determination of ethanol in wine. *Am. J. Enol. Vitic.* **19**: 160-165.

Carroll, R. 1983. Technology and economics of fermentation alcohol -- an update. *Enz. Microb. Technol.* **5**: 103-114.

Chadha, B.S. 1987. Improvement in heat resistance and ethanol tolerance of Saccharomyces cerevisiae strains. M.Sc Thesis. Punjab Agric. Univ. Ludhiana.

Cheema, A. 1989. Production of ethanol from guava. M.Sc. Thesis. Punjab Agric. Univ. Ludhiana.

Chen, H.C. 1990. Non-aseptic, multi- stage, multi- feeding, continuous fermentation of cane molasses to ethanol. *Process Biochem.* **25**(3): 87-92.

Cheryan, M. and Mehaia, M.A. 1984. Ethanol production in a membrane recycle bioreactor. Conversion of glucose using Saccharomyces cerevisiae. *Process Biochem.* **19**(6): 204-208.

*Chung, K.S., Koo, Y.J. and Hwashin, D. 1987. Intergeneric protoplast fusion between Hansenula anomala var. anomala and Saccharomyces cerevisiae. *Korean J. Appl. Microbiol. Bioeng.* **15**: 145-149.

Dallmann, K., Buzas, Z. and Szajani, B. 1987. Continuous fermentation of apple juice by immobilized yeast cells. *Biotechnol. Lett.* **9**(8): 577-580.

D'Amore, T., Celotto, G., Russell, I. and Stewart, G.G. 1989. Selection and optimization of yeast suitable for ethanol production at 40 °C. *Enz. Microb. Technol.* **11**: 411-416.

D' Amore, T., Panchal, C.J., Russell, I. and Stewart, G.G. 1988. Osmotic pressure effects and intracellular accumulation of ethanol in yeast during fermentation. *J. Ind. Microbiol.* **2**(6): 365-372.

D'Amore, T., Panchal, C.J. and Stewart, G.G. 1987. The effect of osmotic pressure on the intracellular accumulation of ethanol in Saccharomyces cerevisiae during fermentation in wort. *J. Inst. Brew.* **93**(6):472-476.

D'Amore, T., Panchal, C.J. and Stewart, G.G. 1988. Intracellular ethanol accumulation in Saccharomyces cerevisiae during fermentation. *Appl. Environ. Microbiol.* **54**(1): 110-114.

D'Amore, T. and Stewart, G.G. 1987. Ethanol tolerance of yeast. *Enz. Microb. Technol.* **9**(6): 322-330.

Davis, B. 1985. Factors influencing protoplast isolation. In: Fungal Protoplasts. J.F. Peberdy and L. Ferenczy (eds.) Marcel Dekker, Inc. New York. pp 45.

Deshpande, V., Keskar, S., Mishra, C. and Rao, M. 1986. Direct conversion of cellulose / hemicellulose to ethanol by Neurospora crassa. *Enz. Microb. Technol.* **8**(3):149-152.

Deutch, C.E. and Parry, J.M. 1974. Spheroplast formation in yeast during the transition from exponential phase to stationary phase. *J. Gen. Microbiol.* **80**: 259-268.

De van Broock, M.R., Sierra, M.F. and de Figueroa, L. 1980. Intergeneric fusion of yeast protoplast. In: Advances in Protoplast Research. G.G Stewart and I. Russell (eds). Vth Int. Symp. on Yeasts, London, Ontario, Canada, 20-25 July, 1980. Pergamon Press Toronto. pp 165-170.

Dickinson, D.P. and Isenberg, I. 1982. Preparation of spheroplasts of Schizosaccharomyces pombe. *J. Gen. Microbiol.* **128**: 651-654.

*Diez-Jerez, M.C. and Mancilha, I. M. de. 1990. Use of sugar beet for obtaining ethanol via fermentation by Zymomonas mobilis CP4 and Saccharomyces uvarum 12 1904 *Arquivos-de-Biologia-e-Technologia.* **33**(4): 915-924.

Dilbaghi, N. 1994. Ethanol production from starch by fusants of Saccharomyces cerevisiae and Saccharomyces diastaticus. M.Sc Thesis. CCS Haryana Agric. Univ. Hissar.

*Dorsemagen, B., Klein, M., Deger, H., Hofer, N. and Wohner, G. 1988. Operational experiences with a large-scale fermenter for continuous production of ethanol. *Dechema Biotechnol. Conferences. No.* **2**:297-307.

*Drawert, F. and Schwank, U. 1986. Use of lactose in mixed substrates for ethanol production. *Branntweinwirtschaft.* **126**(23): 366-370.

*El- Diwany, A.I., El- Abyad, M.S., El- Refai, A.H., Sallam, L.A. and Allam, R.F. 1992. Effect of some fermentation parameters on ethanol production from beet molasses by Saccharomyces cerevisiae Y-7. *Bioresource Technol.* **42**(3): 191-195.

Ernandes, J.R., Matulionis, M., Cruz, S.H., Bertolini, M.C. and Laluce, C. 1990. Isolation of new ethanol-tolerant yeasts for fuel ethanol production from sucrose. *Biotechnol. Lett.* **12**(6): 463-468.

Farahnak, F., Seki, T., Ryu, D.Y. and Ogrydziak, D. 1986. Construction of lactose assimilating and high ethanol producing yeasts by protoplast fusion. *Appl. Environ. Microbiol.* 51: 362-367.

Ferenczy, L., Kevei, F., Szegedi, M., Franko, A. and Rojik, I. 1976. Factors affecting high-frequency fungal protoplast fusion. *Experientia.* **32**: 1156.

*Fernandez, W.L., Barraquio, V.L. and Bernadino, M.H.N. 1981. Alcohol production by eight yeasts in whey. *NSDB Technol. J.* **6**: (4): 19-20.

Figueroa, L.I. de, Cabada, M.A. de and van Broock, M.R. de. 1985. Alcoholic fermentation of starch containing media using yeast protoplast fusion products. *Biotechnol. Lett.* **7**(11): 837-840.

Figueroa, de L.T., Richard, M.F de and van Broock M.R.de. 1984. Interspecific protoplast fusion of the baker's yeast Saccharomyces cerevisiae and Saccharomyces diastaticus. *Biotechnol. Lett.* **6**: 269-274.

Fleet, G.H., Heard, G.M. and Gao, C. 1989. The effect of temperature on growth and ethanol tolerance of yeasts during wine fermentation. VII Int. Symp. on Yeasts, Perugia (Italy). *Yeast.* **2**: 543-546.

Freitas de Morais, S.M., Aquarone, E., Rose, A.H. and Beavan, M.J. 1986. Ethanol effect on fermentative activity from Saccharomyces cerevisiae. *Rev. Microbiol.* **17**(4): 371-375.

Fumi, M.D., Trioli, G., Colombi, M.G. and Colagrande, O. 1988. Immobilization of Saccharomyces cerevisiae in calcium alginate gel and its application to bottle- fermented sparkling wine production. *Am. J. Enol. Vitic.* **39**(4): 267-272

Ghareib, M., Youssef, K.A. and Khalil, A.A. 1988. Ethanol tolerance of Saccharomyces cerevisiae and its relationship to lipid content and composition. *Folia Microbiol.* **33**(6): 447-452.

*Giaja, J. 1919. Emploi des ferments dans les etudes de physiologie cellulaire: Le globule de levure depouille de sa membrane. *C.R. Soc. Biol. Fil. Paris.* **82**: 719-720. (Cited by Davis, 1985).

*Gibbons, W.R. and Westby, C.A. 1989. Cofermentation of sweet sorghum juice and grain for production of fuel ethanol and distillers, wet grain. *Biomass.* **18**(1): 43-47.

Girbes, T. and Parrilla, B. 1986. Correlation between ethanol resistance and incorporation of nutrients by must fermenting yeast. *Cell Mol. Biol.* **32**(5): 545-549.

Gokhale, D.V., Rao, B.S. and Sivarama krishnan, S. 1986. Alcohol dehydrogenase and invertase activities in ethanol tolerant yeasts. *Enz. Microb. Technol.* **8**(10): 623-626.

Gokhale, D.V., Sivaraman, H., Patil, S.G. and Sivaraman, C. 1983. Continuous ethanol production from sugar-cane molasses by immobilized cells of an ethanol tolerant yeast. VIIIth Int. Specialized Symp on Yeast. Bombay. Jan 24-28, 1983. Abstracts of papers. *Indian J. Microbiol.* **23**: 20.

Goma, G., Strehaiano, P., Uribelarrea, J.L., Mota, M., Durand, G. 1983. Kinetic considerations on ethanol production. In: Energy from Biomass. 2nd E.C. Conference. A. Strub, P. Chartier, G. Schleser (eds)., Commission of the European Communities, Luxembourg. Publ. Appl. Sci. Publishers, London (U.K.). pp 1027-1032.

Goyal, R.B. 1989. Comparative ethanol production using Saccharomyces cerevisiae and Zymomonas mobilis. M.Sc. Thesis. Punjab Agric. Univ. Ludhiana.

Goyal, S. 1985. Improvement of ethanol tolerance and flocculation in Saccharomyces cerevisiae. M.Sc. Thesis. Punjab Agric. Univ. Ludhiana.

Goyal, N. and Tauro, P. 1983. Sugar beet as a raw material for bioethanol production in India. VIIIth Int. Specialized Symp. on Yeast. Bombay. Jan 24-28, 1983. Abstr. of papers. *Indian J. Microbiol.* **23**: 25.

Groot, W.J., Waldram, R.H., vander Lans, R.G.J.M. and Luyben, K.C.A.M. 1992. The effect of repeated temperature shock on baker's yeast. *Appl. Microbiol. Biotechnol.* **37**(3): 396-398.

Grote, W. and Rogers, P.L. 1985. Ethanol production from sucrose based raw materials using immobilized cells of Zymomonas mobilis. *Biomass.* **8**(3):169-184.

*Grover, R. 1992. Protoplast production from Endomycopsis fibuligera and fusion with Saccharomyces cerevisiae. M.Sc. Thesis. CCS Haryana Agric. Univ. Hisar.

*Gu, Q.F. 1992. Exploration on the industrial application of Zymomonas mobilis for alcohol production. *Chemical Reaction Engg. Technol.* **8**(4): 399-406.

Gunge, N. and Tamaru, A. 1978. Genetic analysis of products of protoplast fusion in Saccharomyces cerevisiae. *Japan. J. Genet.* **53**(1): 41-49.

Hamlyn, P.F., Bradshaw, R.E., Mellon, F.M., Santiago, C.M., Wilson, J.M. and Peberdy J.F. 1981. Efficient protoplast isolation from fungi using commercial enzymes. *Enz. Microb. Technol.* **3**: 321-325.

Hang, Y.D., Lee, C.Y. and Woodams, E.F. 1986. Solid state fermentation of grape pomace for ethanol production. *Biotechnol. Lett.* **8**(1):53-56.

Hang, Y.D., Lee, C.Y., Woodams, E.F. and Colley, H.J. 1981. Production of alcohol from fruit waste apple pomace. *Appl. Environ. Microbiol.* **42**(6):1128-1129.

Hariantono, J., Yokota, A., Takao, S. and Tomita, F. 1991. Ethanol production from raw starch by simultaneous fermentation using Schizosaccharomyces pombe and a raw starch saccharifying enzyme from Corticium rolfsii. *J. Ferm. Bioeng.* **71**(5): 367-369.

*Heard, G.M. and Fleet, G.H. 1986. Occurrence and growth of yeast species during fermentation of some Australian wines. *Food Technol. Aust.* **38**(1): 22-25.

Heslot, H. 1983. Yeast cloning vectors. VIIIth Int. Specialized Symp. on Yeast. Bombay. Jan 24-28, 1983. Abstr. of papers. *Indian J. Microbiol.* **23**: 35.

Hoffmann, M., Zimmerman, M. and Emies, C.C. 1987. Orthogonal field alternation gel electrophoresis (OFAGE) as a means for the analysis of somatic hybrids obtained by protoplast fusion of different Saccharomyces strains. *Curr. Genet.* **11**: 599-603.

Hoppe, G.K. and Hansford, G.S. 1984. The effect of micro-aerobic conditions on continuous ethanol production by Saccharomyces cerevisiae. *Biotechnol. Lett.* **6**(10):681-686.

*Hospodke, J. 1968. Theoretical and methodological basis of continuous culture of microorganism. *Czechoslovakia. Acad. Sci. Prague* (Cited by Dahiya et al., 1980)

Housset, P., Nagy, M. and Schwencke, J. 1975. Protoplasts of Schizosaccharomyces pombe: An improved method for their preparation and the study of their guanine uptake. *J. Gen. Microbiol.* **90**: 260-264.

*Hsie, M.C. 1993. Isolation and selection of thermotolerant yeast with high yield of ethanol from molasses. Report of the Taiwan Sugar Research Institute. No. **140**: 29-38.

Hunter, K., and Rose, A.H. 1971. Yeast lipid membranes. In: The Yeasts. A.H. Rose and J.H. Harrison (eds.). Academia Inc. London. **2**:211-270.

Ingram, L. O. and Buttke, T.M. 1984. Effects of alcohols on microoganisms. *Adv. Microb. Physiol.* **25**: 253-300.

Isabel, Sa-C, and van Uden, N. 1983. Temperature profiles of ethanol tolerance: effects of ethanol on the minimum and maximum temperatures for growth of yeasts Saccharomyces cerevisiae and Kluyveromyces fragilis. *Biotechnol. Bioeng.* **25**(6): 1865-1867.

Isabel, Sa-C. and van Uden, N. 1986. Ethanol- induced death of Saccharomyces cerevisiae at low and intermediate growth temperatures. *Biotechnol. Bioeng.* **28**(2): 301-303.

Ismail, A.A. and Ali, A.M.M. 1971a. Selection of high ethanol-yielding Saccharomyces. 1. Ethanol tolerance and the effect of training in Saccharomyces cerevisiae Hansen. *Folia Microbiol.* **16**: 346-349.

Ismail, A.A. and Ali, A.M.M. 1971b-Selection of high ethanol-yielding Saccharomyces. II. Genetics of ethanol tolerance. *Folia Microbiol.* **16**: 350-354.

Janderova, B., Cvrckova, F. and Bendova, O. 1990. Construction of the dextrin degrading POF brewing yeast by protoplast fusion. *J. Basic Microbiol.* **39**: 499-505.

*Janderova, S., Puta, F. Bendova, O. 1989. Utilization of yeasts producing amylolytic enzymes and preparation of new strains. *Kvasny Prum.* **35**(7): 208-210.

Jimenez, J. and Benitez, T. 1987. Adaptation of yeast cell membranes to ethanol. *Appl. Environ. Microbiol.* **53**(5): 1196-1198.

Jimenez, J. and Benitez, T. 1988. Selection of ethanol tolerant yeast hybrids in pH-regulated continuous culture. *Appl. Environ. Microbiol.* **54**(4): 917-922.

Jimenez, J. and van Uden, N. 1985. Use of extracellular acidification for the rapid testing of ethanol tolerance in yeasts. *Biotechnol. Bioeng.* **27**(11): 1596-1598.

*Jirku, V. 1987. Ethanol tolerance of yeast cell. *Kvasny Prum.* 33 (4): 106-108.

Jones, R.P. and Greenfield, P.F. 1984. Kinetics of apparent cell death in yeast induced by ethanol. *Biotechnol. Lett.* **6**(7): 471-476.

Jones, R.P. and Greenfield, P.F. 1985. Replicative inactivation and metabolic inhibition in yeast ethanol fermentations. *Biotechnol. Lett.* **7**(4): 223-228.

Jones R.P. and Greenfield, P.F. 1987. Specific and non-specific inhibitory effects of ethanol on yeast growth. *Enz. Microb. Technol.* **9**(6): 334-338.

Kaeppeli, O. and Sonnleitner, B. 1986. Regulation of sugar metabolism in Saccharomyces type yeast. Experimental and conceptual considerations. CRC. *Crit. Rev. Biotechnol.* **4**(3):299-326.

Karanth, N.G. 1983. Some aspects of continuous production of fermentative ethanol. VIIIth Int. Specialized Symp. on Yeast. Bombay Jan. 24-28, 1983. Abstracts of papers. *Indian J. Microbiol.* 23: 8.

Kaur, J. 1992. Processing of substandard fruits for ethanol production. M.Sc Thesis. Punjab Agric. Univ., Ludhiana.

Kavanagh, K. and Whittaker, P. 1990. Formation of spheroplasts and protoplasts in the xylose-fermenting yeast Pachysolen tannophilus. Biotechnol. *Appl. Biochem.* **12**: 57-62.

*Kavanagh, K. and Whittaker, P.A. 1991. Effect of buffer viscosity on the rate of protoplast formation in yeasts. *Biomed. Lett.* **46**(181):57-62.

Kida, K., Kume, K., Morimura, S. and Sonoda, Y. 1992. Repeated batch fermentation process using a thermotolerant flocculating yeast constructed by protoplast fusion. *J. Ferm. Bioeng.* **74**(3): 169-173.

Kida, K., Morimura, S., Kume, K., Suruga, K. and Sonoda, Y. 1991. Repeated batch ethanol fermentation by a flocculating yeast, Saccharomyces cerevisiae IR-2. *J. Ferm. Bioeng.* **71**(5): 340-344.

*Kishkovskaya, S.A. and Bur'yan, N. I. 1980. Thermotolerant variants of wine yeasts. *Mikrobiologiya.* **49**(1): 157-159.

Kleinans, F.W., Lee, N.D., Bard, M., Haak, R.A. and Woods, R.A. 1979. *Chemistry and Physics of Lipids.* **23**: 143 (cited by Ingram and Buttke, 1984)

Kopecka, M. 1975. The isolation of protoplasts of the fission yeast Schizosaccharomyces by Trichoderma viride and snail enzymes. *Folia Microbiol.* **20**: 273-276.

*Krauel, H.H., Menzel, G., Miklos, I. and Sipiczki, M. 1990. Ethanol production by recombinant yeast strain in a fed batch/ batch culture. In: 'Proceedings of the IVth Trilateral Conference on Yeasts'. *Zentralblatt fur Mikrobiologie.* **145**(5): 250.

Krouwel, P.G. and Braber, L. 1979. Ethanol production by yeast at supraoptimal temperatures. *Biotechnol. Lett.* **1**(10): 403-408.

Kumar, U.S., Nagarajan, L., Rehana, F. and Nand, K. 1990. The effect of ethanol on cell wall antigens of Saccharomyces cerevisiae and specific isolation of high ethanol producing strains of this yeast, making use of a serological technique. Antonie van Leeuwen hoek. **58**(1):57-66.

Kumari, J.A. and Panda, T. 1992. Studies on critical analysis of factors influencing improved production of protoplasts from Trichoderma reesei mycelium. *Enz. Microb. Technol.* **14**: 241-248.

Laluce, C., Abud, C.L., Greenhalf, W. and Sanches-Peres, M.F. 1993. Thermotolerance behaviour in sugar cane syrup fermentations of wild type yeast strains selected under pressures of temperature, high sugar and added ethanol. *Biotechnol. Lett.* **15**(6): 609-614.

Laluce, C., Bertolini, M.C., Ernandes, J.R., Spencer, D.M. and Spencer, J.F.T. 1989. Protoplast fusion and thermotolerance of amylolytic yeasts. In: Yeast as a main protagonist of Biotechnology. A Martini, A. Vaughan Martini (eds.). pp S109-S111. Yeast. 2.

Laluce, C., Palmieri, M.C., and da Cruz, R.C.L. 1991. Growth and fermentation characteristics of new selected strains of Saccharomyces at high temperatures and high cell densities. *Biotechnol. Bioeng.* **37**(6): 528-536.

Laplace, J.M., Delgenes, J. P., Moletta, R. and Navarro, J.M. 1991. Combined alcoholic fermentation of D-xylose and D-glucose by four selected microbial strains: process considerations in relation to ethanol tolerance. *Biotechnol. Lett.* **13**(6): 445-450.

*Laude-Bousquet, A. 1989. Process and apparatus for thermal control of wine making. US Patent. 4,814,189.

Leao, C. and van Uden, N. 1982. Effects of ethanol and other alkanols on the kinetics and the activation parameters of thermal death in Saccharomyces cerevisiae. *Biotechnol. Bioeng.* **24**: 1581-1590.

*Lee, C.Y. and Kime, R.W. 1990. Stabilization of wine with honey and sulphur dioxide. US patent. 4,900,564.

Lee, J.H., Williamson, D. and Sch. Roger, P.L.C. 1980. The effect of temperature on kinetics of ethanol production by Saccharomyces cerevisiae. *Biotechnol. Lett.* **2**: 141.

Leite, S.G.F. and Franca, F.P. 1988. Preliminary study of effect of the addition of ethanol to the alcoholic fermentation carried out by Saccharomyces cerevisiae F 1. *Rev. Microbiol.* **19**(4): 430-431.

*Lembke, A., Uderberg, E. and Strobel, H.J. 1990. Process for the production of sparkling wine. US patent 4,948,598.

*Lindeman, L.R. and Rocchiccioli, C. 1979. *Biotechnol. Bioeng.* **21**: 1107. (Cited by Dahiya et al., 1980)

Loureiro-Dias, M.C. and Santos, H. 1990. Effects of ethanol on Saccharomyces cerevisiae as monitored by in vivo 31P and 13C NMR. *Archives of Microbiol.* **153**(4) 384-391.

Luong, J.H.T. 1985. Kinetics of ethanol inhibition in alcohol fermentation. *Biotechnol. Bioeng.* **27**: 280-285.

Maiorella, B.L., Blanch, H.W. and Wilke, C.R. 1984. Feed component inhibition in ethanolic fermentation by Saccharomyces cerevisiae. *Biotechnol. Bioeng.* **26**: (10): 1155-1166.

Melzoch, K., Rychtera, M., Markvichov, N.S., Pospichalova, V. and Basaova, G. 1991. Application of a membrane recycle bioreactor for continuous ethanol production. *Appl. Microbiol. Biotechnol.* **34**(4): 469-472.

Menezes, T.J. B.de, dos Santos, C.L.M. and Azzini, A. 1985. Utilization of bamboo for the production of ethanol. In: Energy from biomass. W.Palz, J. Coombs, D.O.J. Hall (eds.). Elsevier Appl. Sci. Publ. London, U.K. pp. 679-683.

Menzel, G., Schoeps, K., Klein, G. and Haefner, B. 1990. Effect on Saccharomyces cerevisiae of a microbial enzyme preparation with cell wall lytic activities. Acta Biotechnol. **10**(3): 271-275.

Miller, G.L. 1959. Use of dinitrosalicylic acid reagent for determination of reducing sugars. *Anal. Chem.* **31**: 426-428.

Millichip, R.J. 1989. Ethanol production by Zymomonas cultures in yeast-conditioned media. US Patent 4,885,241.

*Minarik, E. 1985. Some microbiological and biotechnological problems in wine making. *Kvasny Prum.* **31**(7-8): 182-183.

Mishra, P. and Prasad, R. 1989. Relationship between ethanol tolerance and fatty acyl composition of Saccharomyces cerevisiae. *Appl. Microbiol. Biotechnol.* 30 (3): 294-298.

Modi, R.I. and Shankar, H.S. 1983. Continuous production of ethanol using Saccharomyces cerevisiae. VIIIth Int. Specialized Symp. on Yeast. Bombay. Jan 24-28, 1983. Abstracts of papers. *Indian J. Microbiol.* **23**: 2.

Moriya, K., Iefuji, H., Shimoi, H., Sato, S.I. and Tadenuma, M. 1988. Ethanol production and flocculation at high temperatures by yeasts bred for fermentation of beet molasses. *J. Brew. Soc. Japan.* **83**(12): 834-837.

Moriya, K., Saito, K., Shimoi, H., Sato, S. and Tadenuma, M. 1987. Conditions of ethanol fermentation. Ethanol fermentation of beet molasses (part 2). J. Brew. Soc. Japan.* **82**(8): 577-581.

Morris, D.W. and Roth, B.A. 1985. Genetic engineering yeast: Methods and application. In Biotechnology: Applications and Research. P.N. Cheremisinoff, R.P. Ouellette (eds.). Publ. Technomic Publishing, Lancaster, P.A. (U.S.A.) pp 172-187.

*Murphy, N.F., Murphy, N.F. de and De-Murphy, N.F. 1988. Ethanol production from blackstrap molasses by Zymomonas mobilis and Saccharomyces sp. *J. Agric. Univ. Puerto-Rico.* **72**(3): 483-484.

Nagodawithana, T.W., Castellano, C. and Steinkraus, K.H. 1974. The effect of dissolved oxygen, temperature, initial cell count and sugar concentration on the viability of Saccharomyces cerevisiae in rapid fermentation. *Appl. Microbiol.* **28**(3): 383-391.

Nagodawithana, T.W. and Steinkraus, K.H. 1976. Influence of the rate of ethanol production and accumulation on the viability of Saccharomyces cerevisiae in "Rapid Fermentation". *Appl. Environ. Microbiol.* **31**(2): 158-162.

Nain, L.R. and Rana, R.S. 1988. Ethanol production from sugarbeet juice by Saccharomyces cerevisiae: Nutrient optimization studies. *J. Maharashtra Agric. Univ.* **13**(2): 141-144.

Nandita, P. 1991. Studies on wines and fuel alcohol production. M.Sc. Thesis. Punjab Agric. Univ. Ludhiana.

Navarro, J.M. and Durand, G. 1978. Alcohol fermentation: effect of temperature on ethanol accumulation within yeast cells. *Ann. Microbiol.* **129B**: 215-224.

Neelam, A. and Amarjit, K.S. 1991. Ethanol production by thermotolerant yeast and its UV resistant mutants. *Acta Microbiol. Pol.* **40**(3/4): 171-175.

*Nemecek, F., Stryakova, K. and Krasny, S. 1990. Application of active dry yeast in technology of sparkling wines. *Kvasny Prum.* **36**(3): 70-72.

Novack, M., Strehaiano, P., Moreno, M. and Goma, G. 1981. Alcoholic fermentation. On the inhibitory effect of ethanol. *Biotechnol. Bioeng.* **23**: 201-211.

Novotny, C., Flieger, M., Panos, J. and Karst, F. 1992. Effect of 5,7- unsaturated sterols on ethanol tolerance in Saccharomyces cerevisiae. *Biotechnol. Appl. Biochem.* **15**(3): 314-320.

Obisanya, M.O., Aina, J.O. and Oguntimein, G.B. 1987. Production of wine from mango using Saccharomyces and Schizosaccharomyces species isolated from palm wine. *J. Appl. Bacteriol.* **63**(3): 191-196.

O'Brien, G. and Whittaker, P. A. 1990. Regeneration of yeast protoplasts prepared using Novozym 234. *Biochem. Soc. Trans.* **18**: 328-329.

*Oderinde, R.A. and Esuosa, K.O. 1988. Batch alcoholic fermentation of molasses: effects of pH, temperature and sugar concentration. *Pak. J. Sci. Ind. Res.* **31**(11):807-810.

Ohta, K. and Hyashida, S. 1983. Role of Tween-80 and monolein in lipid sterol protein complex which enhances ethanol tolerance of sake yeast. *Appl. Environ. Microbiol.* **46**: 821.

Okolo, B., Johnston, J.R. and Berry, D.R. 1987. Toxicity of ethanol, n-butanol and isoamyl alcohol in Saccharomyces cerevisiae when supplied separately and in mixtures. *Biotechnol. Lett.* **9**(6): 431-434.

Ozilgen, M., Celik, M. and Bozoglu, T.F. 1991. Kinetics of spontaneous wine production. *Enz. Microb. Technol.* **13**(3): 252-256.

Panchal, C.J., Bilinski, C.A., Russel, I. and Stewart, G. G. 1986. Yeast stability in the brewing and industrial fermentation ethanol industries. CRC. *Crit. Rev. Biotechnol.* **4**(3): 263-298.

Park, S.C and Baratti, J. 1991. Comparison of ethanol production by Zymomonas mobilis from sugar beet substrates. *Appl. Microbiol. Biotechnol.* **35**(3): 383-391.

Parsons. R.V., Mc Duffie, N.G. and Din, G.A. 1984. pH inhibition of yeast ethanol fermentation in continuous culture. *Biotechnol. Lett.* **6**(10): 677-680.

Patil, S.G., Gokhale, D.V. and Patil, B.G. 1989. Novel supplements enhance the ethanol production in cane molasses fermentation by recycling yeast cells. *Biotechnol. Lett.* **11**(3):213-216.

Patil, S.G. and Patil, B.G. 1990. Acceleration of ethanol production activity of yeast in cane molasses fermentation by the addition of fungal mycelium. *Enz. Microb. Technol.* **12**(2): 141-148.

*Patureau, J.M. 1969. By product of the cane sugar industry. Elsevier, Amsterdam.

Peberdy, J.F. 1979. Fungal protoplasts: isolation, reversion, and fusion. *Ann. Rev. Microbiol.* **33**: 21-39.

Pilando, L.S., Wrolstad, R.E. and Heatherbell, D.A. 1985. Influence of fruit composition, maturity and mold contamination on the color and appearance of strawberry wine. *J. Food Sci.* **50**(4): 1121-1123

Pina, A., Calderon, I. L. and Benitez, T. 1986. Intergeneric hybrids of Saccharomyces cerevisiae and Zygosaccharomyces fermentati obtained by protoplast fusion. *Appl. Environ. Microbiol.* **51**(5): 995-1003.

*Polsinelli, M. 1990. Genetic improvement of wine yeasts. *Industrie delle Bevande.* **19**(110): 495-497, 500.

Pontiveros, C.R., Akontara, J.A. and Rosodro, E.J. 1978. Acid saccharification and alcohol fermentation of unripe banana fruit. *Phillippine J. of Crop. Sci.* **3**(3):153-158.

Porro, D., Martegani, E., Ranzi, B.M. and Alberghina, L. 1992. Lactose/whey utilization and ethanol production by transformed Saccharomyces cerevisiae cells. *Biotechnol. Bioeng.* **39**(8): 799-805.

Prouty, J.L. and Davy, M.K. 1980. The impact of differing raw materials and alternate fermentation methods on process designs for the production of fuel grade ethanol. Proc. of the IVth Int. Symp. on alcohol fuels technology. **1**:57-61.

Raghav, R., Sivaraman, H., Gokhale, D.V. and Seetarama Rao, B. 1989. Ethanolic fermentation of cane molasses by a highly flocculent yeast. *Biotechnol. Lett.* **11**(10): 739-744.

Rao, G. and Mutharasan, R. 1986. Alcohol production by Clostridium acetobutylicum induced by methyl viologen. *Biotechnol. Lett.* **8**(12):893-896.

Reed, G. and Nagodawithana, T.W. 1988. Technology of yeast usage in winemaking. *Am. J. Enol. Vitic.* **39**(1): 83-90.

Retamal, N., Duran, J.M. and Fernandez, J. 1987. Ethanol production by fermentation of fruits and cladodes of prickly pear cactus. *J. Sci. Food Agric.* **40**(3):213-218.

Ribeiro, S.M.G.G., Aboutboul, H., Faria, J.B., Schenberg, A.C.G. and Schmidell, W. 1989. Genetic improvement of Saccharomyces for ethanol production from starch In: Yeast as a main protagonist of Biotechnology. A. Martini, A. Vaughan Martini (eds.). *Yeast* **2**: S11-S15.

Richter, K. 1987. Effects of temperature on microbial ethanol production. II. A thermodynamic interpretation of the temperature profile curve of ethanol production. *Acta Biotechnol.* **7**(2): 127-140.

Richter, K. 1989. A reduced specific ethanol-forming performance of yeast at high biomass concentrations as a result of changed ethanol-tolerance behaviour of the cells under condition of limitation. 1. A theoretical treatment. *Acta Biotechnol.* **9**(1): 17-23.

Richter, K. and Becker, U. 1985. Appearance of inhibition in the ethanol fermentation. 1. Influence of ethanol concentration on the specific rate of ethanol production of Saccharomyces cerevisiae. *Acta Biotechnol.* **5**(1): 9-17.

Richter, K. and Becker, U. 1987. Effects of temperature on microbial ethanol production. 1. The temperature profile curve of ethanol production of yeast strain Saccharomyces cerevisiae Sc. 5. *Acta Biotechnol.* **7**(1):87-92.

Rose, A.H. 1980. Recent research on industrially important strains of Saccharomyces cerevisiae. In: Biology and activities of yeasts. F.A. Skinner, S.N. Passmore and R.R. Davenport (eds.) London Academic Press. pp. 103-109.

Rose, D. 1976. Yeasts for molasses alcohol. Process Biochem. 11: 10-12. Russell, I., Crumplen, C.M., Jones, R.M. and Stewart, G.G. 1986. Efficiency of genetically engineered yeast in production of ethanol from dextrinized cassava starch. *Biotechnol. Lett.* **8**: 169-174.

Saigal, D. and Vishwanathan, L. 1984. Effect of oils and fatty acids on molasses fermentation by distiller's yeast. *Enz. Microb. Technol.* **6**(2): 178-186.

Saito, K., Shimoi, H., Sato, S., Tadenuma, M., Yoshizawa, K., Moriya, K. and Izumi, C. 1990. Yeast strain with high power to produce alcohol by fermentation. United States Patent. US 4910144.

Saito, T., Suzuki, T., Hayashi, A., Honda, H., Taya, M., Iijima, S. and Kobayashi, T. 1990. Expression of a thermostable cellulase gene from a thermophilic anaerobe in Saccharomyces cerevisiae. *J. Ferm. Bioeng.* **69**(5): 282-286.

Sakai, T., Koo, K., Saitoh, K. and Katsuragi, T. 1986. Use of protoplast fusion for the development of rapid starch fermenting strains of Saccharomyces diastaticus. *Agric. Biol. Chem.* **50**(2): 297-306.

Salgueiro, S.P., Sa-Correia, I. and Novais, J.M. 1988. Ethanol-induced leakage in Saccharomyces cerevisiae: kinetics and relationship to yeast ethanol tolerance and alcohol fermentation productivity. *Appl. Environ. Microbiol.* **54**(4): 903-909.

Sathe, S., Shivaraman, H. and Gokhale, D.V. 1992. Protoplast fusion in yeast: Strain improvement in Saccharomyces. *Indian J. Microbiol.* **32**(1): 15-27.

*Savackuk, M. Ya and Tsigankov, P.S. 1971. The effect of temperature on the formation of ethyl alcohol and by-products during the formation of molasses solution. *Kharchova Promstorist.* **13**: 85.

Schwencke, J. and Nagy, M. 1978. Preparation of protoplast of Schizosaccharomyces pombe. *Methods in Cell Biol.* **20**: 101-105.

Seiko, Y., Murakami, T., Kuriyama, H. and Sonoda, Y. 1985. Selection of yeast strains having tolerance to inhibitive conditions. *Rep. Ferment. Res. Inst.* **63**: 55-63.

Shahin, M.M. 1971. Preparation of protoplast from stationary phase cells of Schizosaccharomyces pombe. *Can. J. Genet. Cytol.* **13**: 714-719.

Shahin, M.M. 1972. Relationship between yield of protoplast and growth phase in Saccharomyces. *J. Bacteriol.* **110**(2): 769-771.

Shamala, T.R. and Sreekantiah, K.R. 1988. Use of wheat bran as a nutritive supplement for the production of ethanol by Zymomonas mobilis. *J. Appl. Bacteriol.* **65**(6):433-436.

Sharma, S. and Tauro, P. 1986. Control of ethanol production by Saccharomyces cerevisiae. Proceedings of National Symposium on "Yeast Biotechnology" held at Haryana Agric. Univ., Hisar, India. 12-14 December, 1985.

Singh, B. 1993. Production of wines from Zingiber officinale (Roscoe). M.Sc. Thesis. Punjab Agric. Univ. Ludhiana.

Sipiczki, M. and Ferenczy, L. 1977. Protoplast fusion of Schizosaccharomyces pombe auxotrophic mutants of identical mating-type. *Mol. Gen. Genet.* **151**: 77-81.

Spencer, D.M., Reynolds, N. and Spencer, J.F.T. 1990. Protoplast fusion: application to industrial yeasts. In: Yeast technology. J.F.T. Spencer, D.M. J. Spencer (eds.). pp. 348-354.

Spencer, J.F.T., Spencer, D.M., de Figueroa L., Heluane, H. and Naugues, J.M.1990. Using the kar 1 mutant of Saccharomyces cerevisiae for gene transfer during protoplast fusion. 15th Int. Conference on Yeast Genetics and Molecular Biology. *Yeasts.* **6**: S 599.

Spindler, D.D., Wyman, C.E., Managheghi, A. and Grohmann, K. 1988. Thermotolerant yeast for simultaneous saccharification and fermentation of cellulose to ethanol. *Appl. Biochem. Biotechnol.* **17**:279-293.

Stein, M.A.C.F. 1986. Banana wine. *Cienc. Cult.* **38**(2): 362-366.

Stephen, E.R. and Nasim, A. 1981. Production of protoplasts in different yeasts by Mutanase. *Can. J. Microbiol.* **27**: 550-553.

Stewart, G.G. 1981a. The place of yeast in biotechnology. *Indian J. Microbiol.* **21**(3): 171-210.

Stewart, G.G. 1981b. The genetic manipulation of industrial yeast strains. *Can. J. Microbiol.* **27**: 973-990.

Stokes, S.L. 1970. Influence of temperature on the growth and metabolism of yeasts. In: The Yeasts. A.H. Rose and J. S. Harrison (eds). Volz. Academic Press Inc. London.

Suzuki, H., Hosono, A., Otani, H. and Tokita, F. 1985. Production of ethanol from whey using 'sake' brewing yeasts. *Jap. J. Dairy Fd. Sci.* **34**(4): A85-A90.

Svoboda, A. and Ourednicek, P. 1990. Yeast protoplasts immobilized in alginate: cell wall regeneration and reversion to cells. *Curr. Microbiol.* **20**(5): 335-338.

*Szopa, J.S., Nielepkowicz-Charczuk, A. and Gancarz, M. 1993. New yeast strains for alcoholic fermentation of molasses at higher sugar concentration. Zeszyty Naukowe Politechniki Lodzkiej, *Technologia-i-chemia Spozywcza.* No. **50**: 113-125.

Takuma, S., Nakashima, N., Tantirungkij, M., Kinoshita, S., Okada, H., Seki, T. and Yoshida, T. 1991. Isolation of xylose reductase gene of Pichia stipitis and its expression in Saccharomyces cerevisiae. *Appl. Biochem. Biotechnol.* **28-29**: 327-340.

Tamaki, H. 1986. Genetic analysis of intergeneric hybrids obtained by protoplast fusion in yeast. *Curr. Genet.* **10**: 491-494.

Terrell, S.L., Bernard, A. and Bailey, R.B. 1984. Ethanol from whey: continuous fermentation with a catabolite repression- resistant Saccharomyces cerevisiae mutant. *Appl. Environ. Microbiol.* **48**(3): 577-580.

Tewari, H.K. 1978a. Prospectus of somras making at home. Proc. Production of Somras. Punjab Agric. Univ. **1**: 22-28.

Tewari, H.K. 1978b. Feasibility report on grapes and their processing in Punjab state.

Tewari, H.K. 1978c. Report on the establishment of experimental winery, MARKFED, Chandigarh.

Tewari, H.K. 1978d. Report on the establishment of pilot scale grape processing plant for wines and brandy.

Tewari, H.K. 1979a. Somras. *Yuvrishma.* Punjab Agric. Univ. **6**(2-3): 19.

Tewari, H.K. 1979b. Karele da Somras. *Yuvrishma.* Punjab Agric. Univ. **6**:(4): 21-22.

Tewari, H.K. 1979c. A manual for diabetics. Punjab Agric. Univ. Ludhiana.

Tewari, H.K. 1980. Karele da Somras. *Package Samachar.* Agric. Deptt. Punjab State: 2-4.

Tewari, H.K. 1980. Somras our Vedic drink. Proc. Assoc. Food Sci. Technol. Ludhiana. **3**:15-16.

Tewari, H.K. 1981. Remedy for Madhumeha-Diabetes. *Vigyan Pragati.* **2**:186.

Tewari, H.K. 1986. Karelian da Somras (wine) kiwen banayey. *Changi kheti.* Punjab Agric. Univ. **22**(12):20-21.

Tewari, H.K. and Ghai, S.K. 1978. Grape culture and somras production. Proc. Production of somras. Punjab Agric. Univ. **1**: 8-21.

Tewari, H.K., Grewal, H.S. and Kalra, K.L. 1988. Vinegar production from substandard fruits. *Biological Wastes.* **26**: 9-14.

Tewari, H.K., Grewal, H.S., Marwaha, S.S. and Sharma, R. 1988. Studies on the suitability of Prunus salcina (plum) for vinegar fermentation. Indigenous medicinal plants Symp.: 1-14. Today and Tomorrow's Printers and Publishers, New Delhi.

Tewari, H.K. and Gupta, L.K. 1978. Somras as food and medicines. Proc. Production of Somras. Punjab Agric. Univ. **1**: 28-40.

Tewari, H.K., Marwaha, S.S., Kennedy, J.F. and Rupal, K. 1987. Bioutilization of pineapple waste for ethanol generation. Wood and celluloses, Industrial utilization, Biotechnology, Structure and Properties. Pub. Ellis Horword Ltd. and John Wiley U.K., U.S.A. and Canada. Chap. **27**: 251-259.

Tewari, H.K., Marwaha, S.S., Kennedy, J.F. and Rupal, K. 1989. Bioethanol generation from biopolymers of vegetable wastes. *Int. Ind. Biotechnol.* **9**(5): 15-19.

Tewari, H.K., Marwaha, S.S., Kennedy, J.F. and Singh, L. 1988. Evaluation of acids and cellulose enzyme for the effective hydrolysis of agricultural lignocellulosic residues. *J. Chem. Technol. Biotechnol.* **41**: 261-275.

Tewari, H.K., Marwaha, S.S. and Rupal, K. 1986. Ethanol from banana peels. *Agric. Wastes.* **16**: 135-146.

Tewari, H.K., Marwaha, S.S., Rupal, K. and Singh, L. 1986. Production of ethyl alcohol from banana peels. J. *Res. Punjab Agric. Univ.* 22(4):703-711.

Tewari, H.K., Marwaha, S.S. and Sehgal, N. 1985. Studies on active dry wine yeasts tablets. Proc. National Symp. *Yeast Biotechnol.* Haryana Agric. Univ. Hisar. :153-159.

Tewari, H.K., Marwaha, S.S. and Singh, L. 1988. Screening of yeasts for ethanol production and treatment of dairy industry waste waters. *J. Res. Punjab Agric. Univ.* **25**(1):81-87.

Tewari, H.K., Marwaha, S.S. and Verma, N. 1987. Utilization of substandard pears for the production of wines and vinegar as food and medicine. Punjab Pear Fruit Seminar. Punjab Agric. Univ. Ludhiana.

Tewari, H.K., Sethi, R.P. and Chahal, D.S. 1978. Screening of grape varieties on the Punjab state for their enological qualities. Proc. Industrial Fermentation Symp. Jammu (J&K), India.

Tewari, H.K., Singh, L. and Sethi, R.P. 1982. A note on the utilization of waste potato for the production of ethanol. *The Punjab Hort. J.* **22**(3-4): 214-216.

Thomas, D.S., Hossack, J.A. and Rose, A.H. 1978. Plasma membrane lipid composition and ethanol tolerance in Saccharomyces cerevisiae. *Arch. Microbiol.* **117**: 239-245.

Toda, K., Asakura, T. and Ohtake, H. 1987. Inhibitory effect of ethanol on ethanol fermentation. *J. Gen. Appl. Microbiol.* **33**(5): 421-428.

Tuite, J. 1969. Plant pathological methods- Fungi and bacteria. Burgess Publ. Co., Minneapolis. Minn. USA, 239.

Tyagi, R.D. 1984. Participation of oxygen in ethanol fermentation. *Process Biochem.* **19**(4): 136-141.

Vajpeyi, K., Viswanathan, L. and Agrawal, P.K. 1990. Studies on ethanol production from sugarcane juice by recycled cells of Saccharomyces cerevisiae. Proceedings of the 52nd Annual Convention of the Sugar Technologists' Association of India, Kanpur, India. G71-G80.

van Solingen, P. and van der Plaat, J.B. 1977. Fusion of yeast spheroplasts. *J. Bacteriol.* **130**(2): 946-947.

Verma, G., Sedha, R.K. and Gupta, R.P. 1983a. Rapid ethanol production using cell recycle. VIIIth Int. Specialized Symp. on Yeast. Bombay. Jan 24-28, 1983. Abstract of papers. *Indian J. Microbiol.* 23:13.

Verma, G., Sedha, R.K. and Gupta, R.P. 1983b. Use of yeast cell recycling for rapid ethanol production from molasses. *J. Ferment. Technol.* **61**(5): 527-531.

Viegas, C.A., Rosa. M.F., Sa-Correia, I. and Novais, J.M. 1989. Inhibition of yeast growth by octanoic and decanoic acids produced during ethanolic fermentation. *Appl. Environ. Microbiol.* **55**(1): 21-28.

Viel, G., Bechard, P., Jolicoeur, C. and Beaubien, A. 1987. Toxicity effects of alcohols on Saccharomyces cerevisiae: A Flow microcalorimetry investigation. *Biotechnol. Technol.* **1**(2): 73-78.

Vyas, K. K. and Joshi, V.K. 1982. Plum wine making. Standardization of methodology. *Indian Food Packer.* **36**(6): 80-86.

Walker-Caprioglio, H.M., Rodriguez, R.J. and Parks, L.W. 1985. Recovery of Saccharomyces cerevisiae from ethanol-induced growth inhibition. *Appl. Environ. Microbiol.* **50**(3): 685-689.

Walsh, R.M. and Martin, P.A. 1977. *J. Inst. Brew.* **83**: 169 (Cited by Jones et al., 1981).

Wang, G.S. and Wang, L.H. 1989. Improvement of ethanol tolerance of xylose-fermenting yeast by protoplast fusion. Report of the Taiwan Sugar Research Institute. No. **124**: 29-37.

Wang, Y.W., Song, L.S. and Zhou, Y.L. 1992. Selection of D-xylose and cellobiose-fermenting and ethanol-producing strains by electric field-induced protoplast fusion. *Chinese J. Biotechnol.* **98**(1): 82-86.

Watson, K., Cavicchioli, R. and Dunlop, G. 1984. Primary and secondary heat shock induction of thermal and ethanol tolerance in yeast. Proc. Conv. Inst. Brew. (Aust. NZ sect) (Pub. 1985). **18th**: 229-235.

Weide, H., Nowak, M. and Bergmann, H. 1989. Process for increasing the yield of ethanol in fermentation. German Democratic Republic Patent. DD 265166.

Whalen, P.J. 1988. Development, scale-up and a continuous fermentation process for the cofermentation of cheese whey and corn for ethanol production. *Sciences Engg.* **49**(4): 964-965 (Dissertation Abstr. Int.)

Whalen, P.J., Shahani, K.M. and Lowry, S.R. 1985. A process for cofermentation of whey and corn to produce industrial alcohol. *Biotechnol. Bioeng. Symp.* (**15**): 117-128.

Wilson, J.J. and Ingledew, W.M. 1982. Isolation and characterization of Schwanniomyces alluvius amyloytic enzymes. *Appl. Environ. Microbiol.* **44**: 301-307.

Xia Gao, C. and Fleet, G.H. 1988. The effects of temperature and pH on ethanol tolerance of wine yeasts, Saccharomyces cerevisiae, Candida stellata and Kloeckera apiculata. *J. Appl. Bact.* **65**(5): 405-409.

Yamamoto, M. and Fukui, S. 1977. Fusion of yeast protoplasts. *Agric. Biol. Chem.* **41**(9):1829-1830.

Yamamura, M., Nagami, Y., Vongsuvanlert, V., Kumnuanta, J. and Kamihara, T. 1988. Effects of elevated temperature on growth, respiratory deficient mutation, respiratory activity and ethanol production in yeast. Can. J. Microbiol./ J. Can. Microbiol. **34**(8): 1014-1017.

*Yanase, H. 1993. Strain improvement of the ethanol-producing Zymomonas mobilis by genetic manipulation. *Seibutsu kogaku kaishi.* **71**(3): 179-191.

*Yokomori, Y., Akiyama, H. and Shimizu, K. 1989. Breeding of wine yeasts through protoplast fusion. In: Yeast as a main protagonist of Biotechnology. A. Martini, A. Vaughan Martini (eds.). pp S145-S150. Yeast. 2.

Yusuf, Z., Islam, K.A., Ahmed, A. and Hossain, N. 1993. Bench scale study of ethyl alcohol production from molasses using Saccharomyces cerevisiae. *Chem. Engg. Res. Bull.* **10**: 40-51.

*Zakrezewski, E. and Zmarlicki, S. 1988. Ethanolic fermentation in whey and whey molasses mixtures. II. Two-stage fermentation process of ethanol production from whey and beet molasses. *Milchwissenschaft.* **43**(8):492-496.

Zikmanis, P.B., Kruce, R.V., Auzina, L.P., Margevica, M.V. and Beker, M.J. 1988. Intensification of alcoholic fermentation upon dehydration-rehydration of the yeast Saccharomyces cerevisiae. *Appl. Microbiol. Biotechnol.* **27**(5-6): 507-509.

** Original not seen*

Publisher's Addendum

Publisher's Note

Latin terms and Scientific terms are printed in *italics*.

Author name:
Nandita Paul prior marriage at the time of submitting the dissertation, and **Nandita Gupta** post marriage presently.

Scans of Original (some)

Sample scanned pages from Original for sake of authenticity.

CERTIFICATE I

This is to certify that the dissertation entitled, "Improvements in Efficiency of Yeasts for Ethanol Production", submitted for the degree of Doctor of Philosophy in the subject of Microbiology (Minor: Biochemistry) of the Punjab Agricultural University, Ludhiana, is a bonafide research work carried out by Ms. Nandita Paul [L-91-BS-75-D] under my supervision and that no part of this dissertation has been submitted for any other degree.

The assistance and help received during the course of investigation have been fully acknowledged.

9/8/95

[Dr H K Tewari]

Major Advisor, Mycologist,
Deptt. of Microbiology,
Punjab Agricultural University, Ludhiana

CERTIFICATE II

This is to certify that the dissertation entitled, "Improvements in Efficiency of Yeasts for Ethanol Production", submitted by Ms Nandita Paul [L-91-BS-75-D] to the Punjab Agricultural University, Ludhiana, in partial fulfilment of the requirements for the degree of Doctor of Philosophy in the subject of Microbiology [Minor: Biochemistry] has been approved by the Student's Advisory Committee after an oral examination on the same, in collaboration with an external examiner.

[Dr H K Tewari]
Major Advisor

3.11.95
[External Examiner]
DR. J. K. GUPTA, PROFESSOR,
DEPTT. OF MICROBIOLOGY,
P.U. CHANDIGARH

Head of the Department

4/1/96
Dean, Post-Graduate Studies

Abstract

Eight strains of yeasts, namely *S. cerevisiae*-220, *S. cerevisiae*-360, *S. cerevisiae*-181/3c, *Trichosporon beigelli*, *Candida haemulonii*, *C. parapsilosis*, *Candida* sp.-3(1) and *Candida* sp.-5(2) have been evaluated for their potential to ferment cane sugars to ethanol at three temperatures. *S. cerevisiae*-220 utilized 84.0% of substrate in 60 h at 30 ± 1°C. *S. cerevisiae*-360 fermented substrate more efficiently (69.3%) at 38 ± 1° C. *S. cerevisiae*-220 multiply faster at 30 ± 1 as well as at 38 ± 1°C. *S. cerevisiae*-220 could withstand 13.0% ethanol whereas *S. cerevisiae*-360 tolerated only 8.0% of ethanol supplemented in the growth medium. The protoplasts of *S. cerevisiae*-220 and *S. cerevisiae*-360 fused together could not generate ethanol more efficiently than their parents at 38 ± 1°C. *S. cerevisiae*-220 adapted at 38 ± 1°C utilized 64.4% of substrate during 60 h. Wine with commercially outstanding quality has been prepared from *Zingiber officianale* with *S. cerevisiae*-220-A at 38 ± 1°C.

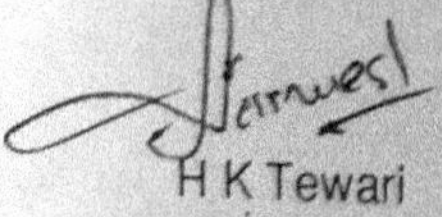

H K Tewari

Nandita Paul

Nandita Paul

CONTENTS

References

A list of some online References.

Béchard, P; Jolicoeur, C; Beaubien, A; Viel, G, 1987
https://link.springer.com/article/10.1007/BF00159325

Caputi, A; Ueda, M; Brown, T, 1968
https://www.ajevonline.org/content/19/3/160

Figueroa, L. I. de; Cabada, M. A. de; Broock, M. R. de van, 1985
https://link.springer.com/article/10.1007/BF01025566

Grote, W; Rogers P L, 1985
https://www.sciencedirect.com/science/article/abs/pii/0144456585900459

Gunge, N; Tamaru, A, 1978
https://www.jstage.jst.go.jp/article/ggs1921/53/1/53_1_41/_pdf

Isabel Sá-Correia; N Van Uden, 1983
https://onlinelibrary.wiley.com/doi/10.1002/bit.260250620

Kopecká, M, 1987
https://pubmed.ncbi.nlm.nih.gov/1100491/

Toda, K; Asakura, T; Ohtake, H, 1987
https://www.jstage.jst.go.jp/article/jgam1955/33/5/33_5_421/_pdf/-char/ja

van Solingen, P; van der Plaat, J B, 1977
https://pubmed.ncbi.nlm.nih.gov/400800/

Walker-Caprioglio; H M, Rodriguez; R J, Parks, L W, 1985
https://pubmed.ncbi.nlm.nih.gov/3907500/

Yamamoto, M; Fukui, S, 1977
https://www.tandfonline.com/doi/pdf/10.1080/00021369.1977.10862772

Zakrzewski, E; Zmarlicki, S, 1989
https://www.osti.gov/etdeweb/biblio/6654620

Epilogue

Life is Precious. Live every moment in Gratefulness, Kindness, Cheerfulness, Joy.

IT IS NOT EASY WITHOUT A MASTER.

सर्वे भवन्तु सुखिनः । सर्वे सन्तु निरामयाः ।

सर्वे भद्राणि पश्यन्तु । मा कश्चिद् दुःख भाग्भवेत् ॥

ॐ शान्तिः शान्तिः शान्तिः ॥

When faith has blossomed in life,
Every step is led by the Divine.

Sri Sri Ravi Shankar

Om Namah Shivaya

जय गुरुदेव

www.ingramcontent.com/pod-product-compliance
Lightning Source LLC
LaVergne TN
LVHW081802240826
846425LV00005B/26

9789395766463